Problems and Proofs in Numbers and Algebra

Richard S. Millman • Peter J. Shiue
Eric Brendan Kahn

Problems and Proofs
in Numbers and Algebra

 Springer

Richard S. Millman
School of Mathematics
Georgia Institute of Technology
Atlanta, GA, USA

Peter J. Shiue
Department of Mathematical Sciences
University of Nevada, Las Vegas
Las Vegas, NV, USA

Eric Brendan Kahn
Department of Mathematics,
    Computer Science and Statistics
Bloomsburg University
Bloomsburg, PA, USA

ISBN 978-3-319-35723-2   ISBN 978-3-319-14427-6 (eBook)
DOI 10.1007/978-3-319-14427-6

Springer Cham Heidelberg New York Dordrecht London

Printed on acid-free paper

Springer International Publishing AG Switzerland is part of Springer Science+Business Media
(www.springer.com)

Preface

The transition from studying calculus or differential equations to learning about proofs is one that is enormously interesting as it shows how exciting mathematics can be. The most important aspect of this stage of the transition for students is the need for rigorous mathematical reasoning. The benefits to readers who are moving from calculus to more abstract mathematics are to develop the ability to understand proofs through Problems and Proofs in Numbers and Algebra (PPNA) on their ways to analysis, abstract algebra, etc. which come next. Our goal is for students to focus on how to both prove theorems and solve problem sets which have depth—multiple steps are needed to prove or solve.

Our approach, "solving rigorous problems and learning how to prove," gives a platform of two specific content themes, number theory and algebra (polynomials), with some aspects of applications. Undergraduate mathematics students will then learn how to prove and solve problems because they are comfortable with the two content themes. Furthermore, our approach is that the content areas of this book allow students to develop a natural and conceptual understanding of the mathematics on their path forward. Students will study the concept with clarity, precision, and a mathematical habit of the mind through these problems. They will gain the foundations of mathematical proof techniques and styles.

The key to the text is its interesting and intriguing problems, exercises, theorems, and proofs, showing how students will transition from the usual, more routine calculus to abstraction while also learning how to "prove" or "solve". Its applications such as RSA cryptosystems, Universal Product Code (UPC), and International Standard Book Number (ISBN) are included in sections of the third chapter.

Problems and proofs are the heart of mathematics. The goal of conceptual understanding grows as a large number of problems and examples that reward curiosity and insightfulness over simplicity. The problems are multi-step and require the reader to think. An intriguing variety of problems range from moderate to thoughtful to deep.

Each problem set begins with a few easy problems. After that, and in coordination with our approach to the subject of depth and conceptual understanding, many of the other problems require multi-step solutions, whether they be problems or proofs. In addition, there are exercises in the text; this difference between exercises, examples, and problem sets is that the exercises stay close to the examples of the section allowing students the immediate opportunity to practice developing techniques.

Furthermore, some problems are motivated from various mathematics competitions. Dr. Shiue's significant experience with problems includes constructing problems and proofs while he is involved in the American Regional Mathematics League (1997–2009), American Mathematical Competitions/MAA (since 1998), Taiwan Regional Mathematics League (since 1997), and Taiwan Junior High School Mathematics Competitions (since 2002). His work with competitions has enriched the quality of our problems.

Some of the concepts include the following:

Number Theory, Algebra, Proofs with approaches to them, Division Algorithm, Euclidean Algorithm, Greatest Common Divisor, Least Common Multiple, the Remainder Theorem, Diophantine Equations and Counting, Equivalence Classes, Divisibility of both Integers and Polynomials, Factoring Polynomials and Roots, Matrices (in the plane and in 3-space), Cramer's Rule, and Determinants, among others.

This text has been revised by the authors over 7 years. An earlier course has been used twice at the University of Kentucky (Problem Solving for Middle School Teachers) and at the University of Nevada, Las Vegas (four times as a secondary resource for future teachers). High school teachers can use the PPNA material in their classroom for strong and advanced students through numbers and algebra.

An advanced math course, Problems and Proof, at the public high school in Georgia, Gwinnett School of Math, Science and Technology (GSMST), has been a basis for PPNA. PPNA has been revised each spring semester from 2011 to 2014 by the authors and four Georgia Institute of Technology graduate students. Justin Boone has done very well as the individual who not only has helped with the typesetting via LATEX but also has given us fine advice. We very much appreciate the advice of Daniel Connelly, S. Greyson Daugherty, and Nolan Leung for their excellence in teaching the PPNA course at GSMST and the help of Scott MacDonald, graduate student at UNLV.

We thank the high school and college students and teachers who worked through the various revisions of the draft, who were a pleasure to collaborate with us, and who are to continue to make mathematics an even more interesting place. We would be happy to receive comments about the book and to respond. Please send to richard.millman@math.gatech.edu, shiue@unlv.nevada.edu, or ekahn@bloomu.edu.

Dr. Millman's granddaughter, Bluma Millman, enjoys a wonderful mathematics major with algebra and number theory and Sandy has helped much. Dr. Shiue would like to thank his wife, Stella, for her full support during the preparation of this book, and it is with great pleasure that Dr. Kahn would like to express his full gratitude to his wife Emily for her support and encouragement throughout all phases of this project.

In addition, Dr. Tian-Xiao He, Illinois Wesleyan University, and Dr. William Speer, UNLV, have reviewed PPNA much and given us good advice. We thank Dr. Derrick DuBose, Chairman, Department of Mathematical Sciences, UNLV, for his support.

Atlanta, GA   Richard S. Millman
Las Vegas, NV   Peter J. Shiue
Bloomsburg, PA   Eric Brendan Kahn

Contents

Part I
The Integers

Chapter 1
Number Concepts, Prime Numbers, and the Division Algorithm

Introduction The topic of this section is the divisibility of integers, the basic building blocks (called prime numbers) for integers, and how to apply this foundation to problem solving in combinatorics and word problems in which integers solutions are sought (Diophantine equations).

Brief Description The operations of addition and multiplication are two ways to combine integers to get a third integer (this is called the closure property). We'll now talk a bit about trying to do the operations in reverse (subtraction and division) and discuss the formal definitions which will give rise to the big question of this half of the text and the big theorem, Fundamental Theorem of Arithmetic, and its applications.

Addition and multiplication satisfy rules that we expect such as associativity, commutativity and the existence of an identity for the operation. More precisely, if a, b and c are any integers then

$$(a + b) + c = a + (b + c) \text{ and } a(bc) = (ab)c \qquad \text{(associativity)}$$

$$a + b = b + a \text{ and } ab = ba \qquad \text{(commutativity)}$$

Furthermore, there is an integer 0 such that

$$a + 0 = 0 + a = a \qquad \text{(0 is the additive identity)}$$

and there is an integer 1 such that

$$a \cdot 1 = 1 \cdot a = a \qquad \text{(1 is the multiplicative identity)}$$

© Springer International Publishing Switzerland 2015
R.S. Millman et al., *Problems and Proofs in Numbers and Algebra*,
DOI 10.1007/978-3-319-14427-6_1

In addition, when we mix these two operations, they respect each other. We have two distribution properties:

$$(a + b)c = ac + bc.$$

$$a(b + c) = ab + ac$$

There is one important difference between these operations and that is whether or not one can "go backwards" from the identity—the existence of an INVERSE. For addition, "go backwards" means that given any integer a, there is another integer x (written as $-a$) such that $a + x = 0$. In more formal language, there is an additive inverse for each integer and that number is actually an integer. We express this fact by saying that the set of integers is CLOSED UNDER TAKING ADDITIVE INVERSES.

For multiplication, "go backwards" would mean that, if a is a given integer, is there an integer x such that a times x equals the multiplicative identity, 1? Unlike addition, "going backwards" rarely works for multiplication. For example, there are no solutions to the equation $ax = 1$ when $a = 0$. That is, $a = 0$ has no multiplicative inverse. Furthermore, when there is a solution to $ax = 1$, that solution is not an integer unless a is 1 or -1.

There is a difference here. While the set of integers *is* closed under addition and taking additive inverses (subtraction), the set of integers is closed under multiplication but *is not* closed under taking multiplicative inverses (division). This last fact can be seen easily through another example. The solution to

$$3x = 1$$

is $x = \frac{1}{3}$ but $\frac{1}{3}$ is not an integer. Other that ± 1, multiplicative inverses don't exist in the set of integers. On the other hand, we may still be able to answer the **question of divisibility** which is the subject of Sect. 1.1.

Question If a and b are integers, is there an *integer* x such that

$$ax = b?$$

The above question is the motivation for the whole of Chap. 1. We will begin by treating the topic of divisibility by asking what is the relationship between integers a and b which ensures you can solve the equation of one variable x, $ax = b$, in such a way that the solution x, is actually an integer?

Going further, when does an integer equation in two variables have a solution which involves only integers and what are those solutions? Said mathematically, if a_1, a_2 and b are integers with $a_1 \neq 0$ and $a_2 \neq 0$, are there integers x and y such that

$$a_1 x + a_2 y = b?$$

Such equations are called **linear Diophantine equations** and are related to the subject of counting as we'll discuss in Chap. 2. We will then define what is meant for elements of a set to be the same, **equivalence classes**, and apply this approach to both **clock arithmetic** and a rigorous definition of **fractions**.

1.1   Beginning Number Concepts and Prime Numbers

Overview We start with $\mathbb{Z}$, the collection of all integers, and $\mathbb{N}$, the collection of all positive integers. The introduction to Chap. 1 discusses the operations of integers and the building blocks of prime numbers. These symbols will be used throughout this text and are common throughout mathematics.

Definition. If a and b are integers and $a \neq 0$, then a DIVIDES b (or a is a DIVISOR of b) if there is an integer c such that

$$ac = b. \tag{1.1.1}$$

In this case, we also say that (i) a is a DIVISOR of b, (ii) a is a FACTOR of b, or (iii) b is a MULTIPLE of a.

Notation. We will write $a|b$ to mean that a divides b or b is DIVISIBLE by a. The symbol $a \nmid b$ means that "a does not divide b." Thus $3|9$ but $4 \nmid 9$.

Remark. When $a \neq 0$, there is always a real number, c, which is a solution to (1.1.1); it is just that the solution is not necessarily an integer. The point of divisibility is that the solution c to (1.1.1) is actually an <u>integer</u>.

Remark. The symbol $a|b$ has nothing (or little) to do with the fraction $\frac{b}{a}$. To say that $3|6$ means that 6 is divisible by 3; that is, there is an integer c with $3c = 6$ (of course, $c = 2$ solves the equation). On the other hand, $3 \nmid 8$ since there is no *integer* c with $3c = 8$. There is a fraction (which is not an integer), $x = \frac{8}{3}$, with $3x = 8$. Since $\frac{8}{3} = 2\frac{2}{3}$ is not an integer, 3 does not divide 8, $3 \nmid 8$.

The definition of divisibility quickly gives the following two properties stated as Exercises 1 and 2 which are used later in the text. The first exercise shows that, if an integer divides each of two numbers, then it divides both their sum and their difference.

Exercise 1. If $a|b$ and $a|c$ then prove that $a|(b + c)$ and $a|(b - c)$.

Exercise 2. Let $a, b, c \in \mathbb{Z}$. Then prove that

A. If a divides b and b divides c, then $a|c$.
B. If a divides b and a divides c, then, for any integers k and l, a divides the integer $kb + lc$.

Just as substances are made up of molecules and molecules are made up of atoms and atoms are made up of electrons, neutrons and protons, so there exist basic building blocks of integers which are called *prime numbers*. We will continue to carefully define terms and present the the Fundamental Theorem of Arithmetic (Theorem 1) which makes the previous sentence mathematically precise. We assume that you have already seen prime numbers and so shall not spend much time on motivating examples.

Definition. If n is an integer and $n > 1$ then n is a PRIME NUMBER if the only positive divisors of n are 1 and n.

We say *n* is a COMPOSITE NUMBER if $n > 1$ and not a prime number.

Suppose we start with a given integer greater than 1, say $n = 315$ and start to look at its factors until we can't go any further. This is the mathematical idea of writing an integer as the product of basic building blocks (prime numbers).

Remark. How many divisors does a prime number, *n*, have? The answer is not two, it's four and they are ± 1 and $\pm n$. The definition of a prime number talks about *positive* divisors so there are no contradictions here (this point is often overlooked by students).

There are 5 ways to factor 315 into 2 numbers (ignoring order). Here are 4 of them (9×35 is the fifth).

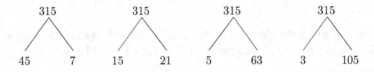

Each of these can be further factored to give rise to the trees (they are actually called "factor trees") below.

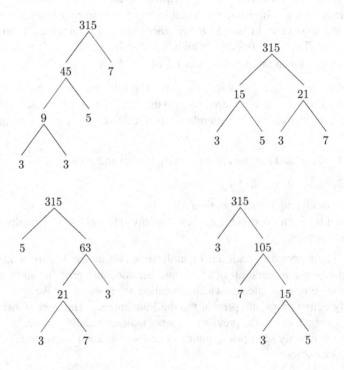

The important principle to be noticed here is that no matter which way we continued to factor (and there are more than we have written), the final result always has two 3's, one 5, and one 7, and each of the integers 3, 5, and 7 are prime numbers. The Fundamental Theorem of Arithmetic is a formal statement of this fact. Its proof requires more advanced material (see References) and so is omitted. It is also referred to as "UNIQUE PRIME FACTORIZATION" because $315 = 3^2 \cdot 5 \cdot 7$.

Theorem 1 (Fundamental Theorem of Arithmetic). *Each integer $n > 1$ can be decomposed as a product of prime numbers (some of which may appear more than once). Furthermore, this decomposition is unique except for the order of the factors.*

The statement of uniqueness means that the prime factors are always the same (except for the order we list them) and the number of times (MULTIPLICITY) a prime appears in the factorization is always the same. The prime factors of 315 are 3, 5, and 7 and 3 appears twice (we say that "3 has multiplicity 2").

By collecting like primes and replacing them by a single factor, we can rephrase Theorem 1 as a corollary:

Corollary 1. *Any positive integer $n > 1$ can be written in exactly one way in the form*

$$n = p_1^{k_1} p_2^{k_2} \ldots p_r^{k_r}$$

where for $i = 1, 2, \ldots, r$ each k_i is a positive integer, called a multiplicity, each p_i is a prime, and $p_1 < p_2 < \ldots < p_r$. We call this representation the PRIME POWER DECOMPOSITION OF n.

At this point, we will assume knowledge of the language used to describe a set but not any formal set theory properties. We also assume the reader is familiar with set notation and the idea that a set is a collection of objects. We will briefly review the notations of union, intersection, and empty and disjoint sets. Note that the concept of a set was first considered by Georg Cantor (1845–1918).

Notation. *We say x is an* ELEMENT *of the set A if x is a member of A and we write* "$x \in A$".

To list a set, we write out all of the elements of the set A inside a set of braces $\{ , \}$ or give a rule which defines the set. In the notation below, we pronounce the symbol "|" as "such that". For example, let A be the set of integers between 1 and 10 inclusive. Then we can write A in two distinct ways:

$$A = \{1, 2, 3, 4, 5, 6, 7, 8, 9, 10\}$$

or

$$A = \{x \in \mathbb{N} \mid 1 \le x \le 10\}.$$

While the symbol "|" when used in a description of a set means "such that", remember that "|" when used with two integers means "divides". Thus $7|21$ (which is true) is a statement and has nothing to do with constructing a set. We can construct new sets from two others (say A and B) in a number of ways.

The most useful constructions are the intersection of A and B (which consists of all elements which are common to A and B) and union of A and B (which consists of those elements which are in A or B).

Definition. The EMPTY SET, written $\emptyset$, is the set with no elements in it.

Definition. If A and B are sets then the INTERSECTION OF A AND B, written $A \cap B$, is given by

$$A \cap B = \{x \mid x \in A \text{ and } x \in B\}.$$

If $A \cap B = \emptyset$ then A and B are DISJOINT.

It is, for example, easy to see that if A is the set of integers which are multiples of 3 and B is the set of integers which are multiples of 2, then $A \cap B$ is the set of integers which are multiples of 6.

Definition. If A and B are sets, the UNION OF A AND B, written $A \cup B$, is given by

$$A \cup B = \{x \mid x \in A \text{ or } x \in B\}$$

Remark. The term "or" in the definition is inclusive. In other words, $x \in A \cup B$ means either x is in A or x is in B (or x is in both A and B.)

Definition. If $n \in \mathbb{Z}$ and n is divisible by 2, then n is called an EVEN INTEGER (or just "even"). If $n \in \mathbb{Z}$ is not even, then n is ODD.

Of course, $n = 0$ is even and so we can represent the sets of even and odd integers respectively as:

$$\text{Even Integers} = \{n \in \mathbb{Z} \mid n = 2k \text{ for some } k \in \mathbb{Z}\},$$

and

$$\text{Odd Integers} = \{m \in \mathbb{Z} \mid m = 2l + 1 \text{ for some } l \in \mathbb{Z}\}.$$

Example 1. Prove that the sum of any two even integers is even.

Solution. Let n and m be even. We must show that $(n + m)$ is twice some integer. Since both are even, by the definition of "even", there are integers ℓ_1 and ℓ_2 with

$$n = 2\ell_1 \text{ and } m = 2\ell_2$$

Thus $n + m = 2\ell_1 + 2\ell_2 = 2(\ell_1 + \ell_2)$ and so $n + m$ is twice the integer $\ell_1 + \ell_2$ which is the definition of $n + m$ being even.   //

Exercise 3. Prove that the sum of any two odd integers is even.

One useful observation is that breaking proofs up into a couple of different cases can really help simplify the work. Suppose that we have a statement in equation form about x and y with at least one of them, say x, taking only integer values. One way to break up the proof is to split the possibilities for x into two cases, one for even numbers and the other for odd. For example, since no two consecutive integers can both be odd, we see that the product of two consecutive integers must be divisible by two. Can you write a formal proof of this statement (Problem 3 of Problem Set 1.1)

Example 2. Find all integers n such that $n^2 - 2n - 3$ is prime.

Solution. The method here is to factor the given polynomial and since the result is a prime number, one of the two factors must be ± 1. We recognize that we can factor the expression and write

$$n^2 - 2n - 3 = (n - 3)(n + 1).$$

Since n is an integer then both $n - 3$ and $n + 1$ are also integers. Thus, the key is if we want $n^2 - 2n - 3$ to be a prime number then either $n - 3 = \pm 1$ or $n + 1 = \pm 1$. Thus, n can be $4, 2, 0$, or -2. We must still check to see if these values give prime numbers:

$$4^2 - 2 * 4 - 3 = 5$$
$$2^2 - 2 * 2 - 3 = -3$$
$$0^2 - 2 * 0 - 3 = -3.$$
$$(-2)^2 - 2 * (-2) - 3 = 5$$

Since prime numbers must greater than 1, the only solutions are $n = 4$ and $n = -2$.   //

Example 3. Find all $x \in \mathbb{Z}$ such that $x^4 + 4$ is a prime.

Solution. Trying to factor $x^4 + 4$ isn't at all obvious! There are two steps needed to perform the factorization. One is to express $x^4 + 4$ as the difference of two squares and then factor. The idea is to do some algebra first by adding $(4x^2 - 4x^2)$ (which is really zero, of course) to the given expression and then factoring using the "difference of squares" formula, which says that for w and z

$$w^2 - z^2 = (w - z)(w + z).$$

In our case,

$$x^4 + 4 = x^4 + 4x^2 + 4 - 4x^2 = (x^2 + 2)^2 - (2x)^2.$$

The difference of squares formula then gives

$$x^4 + 4 = [(x^2 + 2) - 2x][(x^2 + 2) + 2x]$$

so that

$$x^4 + 4 = (x^2 - 2x + 2)(x^2 + 2x + 2).$$

The point is that since we want to know for what x the left hand side is prime, we can look for the values of x where exactly one of the terms on the right hand side is either $+1$ or -1. So we are left to find solutions of the four quadratic equations:

$$x^2 - 2x + 2 = 1, \, x^2 - 2x + 2 = -1$$
$$x^2 + 2x + 2 = 1, \, x^2 + 2x + 2 = -1.$$

The second and fourth of these equations yield no integer solutions by the quadratic formula. However the first and third equations can be rewritten as:

$$(x - 1)^2 = 0 \text{ and } (x + 1)^2 = 0.$$

So we have two possible x values that will make $x^4 + 4$ prime, namely 1 and -1. We see if we plug either answer in for x, then the calculation yields 5 which is prime. The only x that makes $x^4 + 4$ prime is when $x = \pm 1$.   //

Let's summarize the technique for solving a problem of the kind "find the values of x such that the polynomial $p(x)$ is a prime number". We are motivated by the definition of a prime number and so we first do some algebra to get the expression $p(x)$ to be the product of two others, then we set each of those two expressions to 1 and then -1. We thus have a list of all possible solutions to the problem and we need to check to see which x values actually give a prime number as an answer. This type of problem can be deceptively complex as sometimes very simple expressions have difficult answers. For example, are there infinitely many primes of the form $a^2 + 1$ with $a \in \mathbb{Z}$? Number theorists believe that there are infinitely many. (See [4] for more information.)

Exercise 4. Find all $x \in \mathbb{Z}$ so that

$$x^4 - 6x^2 + 25$$

is a prime number. (Hint: Show that this expression is the same as $(x^2 + 5)^2 - 16x^2$. Can you factor the latter in a manner similar to Example 3?)

We shall close this section by proving two more facts about primes by using the concepts of this section.

Theorem 2 (Euclid's Theorem). *There are infinitely many primes.*

Proof. Assume that there are only finitely many primes $p_1, p_2, \ldots, p_n$. Let $m = p_1 p_2 \cdots p_n + 1$. We know from the Fundamental Theorem of Arithmetic that there is a prime p such that $p|m$. Since p is a prime and $p_1, p_2, \ldots, p_n$ are *all* of the primes, there must be a j such that $p = p_j$ ($j = 1, 2, \cdots, n$). Now $p|p_1 p_2 \cdots p_n$ and $p|m$. By Exercise 1, we have $p|1$. This is a contradiction. ■

Proposition 1. *If n is a composite number, then n has a prime divisor p such that $p \leq \sqrt{n}$.*

Proof. By hypothesis, $n = ab$ where $2 \leq a \leq b < n$. Therefore, $n \geq a^2$ or $a \leq \sqrt{n}$. Let p be a prime such that $p|a$. Such a p exists by the Fundamental Theorem of Arithmetic. Thus, $p|n$ and $p \leq a \leq \sqrt{n}$. ■

For example, 40 is a composite, $2|40$, $2 \leq \sqrt{40}$ and $5|40$, $5 \leq \sqrt{40}$. Following Proposition 1, we can conclude that to test whether the number n is prime, it is not necessary to test whether n has a prime divisor greater than $\sqrt{n}$.

For example, is $n = 311$ a prime? The primes less than $\sqrt{311} < 18$ are $2, 3, 5, 7, 11, 13, 17$. Since none of these primes are a divisor of 311, we conclude that 311 is prime.

Problem Set 1.1:

1. (a) List five of the elements of the set

$$\mathcal{O} = \{m \in \mathbb{Z} \mid m = 2\ell - 1 \text{ for some } \ell \in \mathbb{Z}\}.$$

   (b) How would you describe the set $\mathcal{O}$ in words?
   (c) Look at the definition of the set of odd numbers in this text. Is there a contradiction between that definition and the defining properties of $\mathcal{O}$?
2. Prove that the sum of an odd integer and an even integer is odd.
3. Prove that the product of two consecutive integers is even.
4. Prove that the only even integer that is prime is the number 2.
5. Find positive integers, p and q, both prime, such that $p + q = 313$.
6. If n is an integer, prove the sum of its square and the square of the integer two more than n is even.
7. (a) Suppose that $x, y \in \mathbb{Z}$ and both are even or both are odd. Prove the difference of their squares $(x^2 - y^2)$ is divisible by 4.
   (b) If $x, y \in \mathbb{Z}$, x is odd, and y is even, is the same result as 7a true?
8. Prove that the sum of the squares of any two consecutive integers is odd.
9. If $3 \nmid n$, then prove that $(n^2 - 1)$ is divisible by 3.
10. If a and b are odd numbers, then prove that $a^3 - b^3$ is even.
11. Find all $x \in \mathbb{Z}$ such that $x^2 - 4x - 5$ is prime.
12. Find all $x \in \mathbb{Z}$ such that $x^2 - 2x - 15$ is prime.

13. Show that there are no odd integers x greater than 1 for which $x^2 + 1$ is a prime number.

14. Find all $x \in \mathbb{Z}$ such that $x^4 + x^2 + 1$ is a prime if there are any.

15. Find all $x \in \mathbb{Z}$ (or show that none exist) such that $x^4 + 3x^2 + 4$ is a prime.

16. Find all $x \in \mathbb{Z}$ (or show that none exist) such that $x^4 - 8x^2 + 16$ is a prime.

17. Find the smallest positive integer n such that

$$p(n) = n^2 + n + 17$$

     is prime.

18. A number of the form $2^n - 1$ for $n \in \mathbb{N} - \{1\}$ is called a Mersenne number (named after the French monk, Father M. Mersenne (1588–1648) who did much to stimulate mathematical research). It is easy to check that $2^2 - 1 = 3$ and $2^3 - 1 = 7$ which are primes. However $2^{11} - 1 = 2047 = 23 \times 89$ which is not prime. A prime number of the form $2^n - 1$ is called a Mersenne prime. It is unknown whether or not there are infinitely many Mersenne primes, the largest known to date is $2^{57,885,161} - 1$ (See Wikipedia.com as of 2013). Show that $2^n - 1$ is a composite number if n is a composite number.

19. Let $f(x) = x^2 + x + 41$. Show that $f(n)$ is prime for $0 \le n \le 39$, but $f(40)$ is a composite number.

20. By using Proposition 1, determine which of the following integers are prime.

     (a) 403

     (b) 157

     (c) 1999

1.2   Divisibility of Some Combinations of Integers

Overview The first two paragraphs of this section consist of two seemingly different kinds of results. The first type is to state general results (such as "If an integer divides two others then it divides their sum"). The second sort is to ask the reader to find all integers which satisfy certain divisibility criteria ("If n is a non-negative integer, find all values for n such that $(3n - 8)$ divides $(2n + 3)$, which is Example 8).

In fact these two types of problems are similar as they both start with using a variable name (usually n) and asking for what n is a condition true. For the first kind of problem, the answer is "all n" (or perhaps "all odd n", "all even n", etc.). For the second, the solution will be a few (or more) values of n for which the criterion is true. Perhaps more importantly, the first kind of result (such as "The product of two consecutive numbers is even".) can be used to solve both types of problems.

The next set of examples and exercises can be approached by giving variable names to the integers involved and doing a little bit of algebra. It is useful to remember that if we are dealing with an odd integer n, then n may be written as $2k + 1$ for some $k \in \mathbb{Z}$. In Example 1, we used a similar technique for a result about the sum of even integers.

Example 4. Show that the sum of 3 consecutive integers is always divisible by three.

Solution. Let $n \in \mathbb{Z}$. The three consecutive integers are then n, $n + 1$, and $n + 2$ and so the sum of three consecutive integers, called s is

$$s = n + (n + 1) + (n + 2) = 3n + 3 = 3(n + 1).$$

Thus s is 3 times an integer and so $3 | s$.   //

Note One can also let the three consecutive integers be $n - 1, n$, and $n + 1$.

The generalization of Example 4 is interesting. See Problem 1 of the Problem Set 1.2.

Example 5. If a and b are odd, then prove that $a^2 - b^2$ is divisible by 8.

Solution. Since a and b are both odd, there are integers k and l with $a = 2k + 1$ and $b = 2l + 1$. Thus

$$
\begin{aligned}
a^2 - b^2 &= (2k + 1)^2 - (2l + 1)^2 \\
&= 4k^2 + 4k - 4l^2 - 4l \\
&= 4(k^2 + k - l^2 - l).
\end{aligned}
$$

Thus $a^2 - b^2$ is divisible by 4 but that is not enough to also say it is divisible by 8. Factoring the above equation gives:

$$a^2 - b^2 = 4(k(k + 1)) - 4(l(l + 1)). \qquad (1.2.1)$$

Now using the insight that $k(k+1)$ and $l(l+1)$ are consecutive integers and so they must both be even (divisible by 2), we can write $k(k + 1) = 2r$ and $l(l + 1) = 2s$ with $r, s \in \mathbb{Z}$. Substituting into Eq. (1.2.1),

$$a^2 - b^2 = 8r - 8s = 8(r - s)$$

and so $a^2 - b^2$ is divisible by 8.   //

We believe strongly that **multiple steps** in a problem is most important. This example requires multiple steps: understanding the definition of odd, doing some algebra, seeing that we are not yet finished (divisible by 4 at this stage), more algebra, starring the equation to see that we are multiplying two consecutive integers (and so have an even number), and then finishing the problem by combining terms. This is the typical method of problem solving—multiple layers with a number of insights, separated by some routine (or clever) computations! A problem where we just write out an answer doesn't give insights but multiple levels do.

Example 6. Let $n \in \mathbb{N}$ and suppose that n is odd. Prove that

$$32 \mid (n^2 + 11)(n^2 + 15).$$

Solution. Since n is odd, it is of the form $n = 2l + 1$ for some $l \in \mathbb{Z}$. The integer on the right-hand of the assertion becomes

$$
\begin{aligned}
(n^2 + 11)(n^2 + 15) &= ((2l + 1)^2 + 11)((2l + 1)^2 + 15) \\
&= (4l^2 + 4l + 12)(4l^2 + 4l + 16) \\
&= 16(l^2 + l + 3)(l^2 + l + 4).
\end{aligned}
$$

The second two numbers on the right-hand side are consecutive integers. By Problem 3 of Problem Set 1.1, their product must be even, say written as $2k$, where k is an integer. Thus, since we have k defined by the equation

$$(l^2 + l + 3)(l^2 + l + 4) = 2k,$$

we see that

$$(n^2 + 11)(n^2 + 15) = 16(2k) = 32k$$

and so $32 \mid (n^2 + 11)(n^2 + 15)$.   //

Exercise 5. Let $n \in \mathbb{N}$ and suppose that n is odd. Prove that

$$32|(n^2 + 4n - 1)(n^2 + 4n + 3).$$

Exercise 6. Let $n \in \mathbb{N}$ and suppose n is odd. Prove that

$$32|(n^2 + 3)(n^2 + 7).$$

We now use Exercises 1 and 2 to solve some divisibility problems.

Example 7. If a, b and c are integers such that $a|(b + c)$ and $a|(b - c)$, then $a|2b$ and $a|2c$.

Solution. Exercise 1 shows that a must divide both the sum of $(b + c)$ and $(b - c)$ and their difference. Therefore $a|2b$ and $a|2c$.   //

Example 8 is the first of the problems that uses Exercises 1 and 2 (if a divides two integers, then it divides both their sum and their difference). It is an extremely useful technique when we remember that any integer divides itself! Thus if $3n - 8$ is an integer then $(3n - 8)|(3n - 8)$ and so $(3n - 8)$ divides any multiple of $(3n - 8)$, so in particular, $(3n - 8)|(-2)(3n - 8)$. This very simple (but clever) observation lets us solve Example 8 because we end up with $(3n - 8)$ dividing an integer (25) which does not contain n as a factor.

Example 8. Find all values of n such that both n and $\dfrac{2n+3}{3n-8}$ are positive integers.

Solution. Since any integer divides itself, $(3n-8)|(3n-8)$ and the example says that $(3n-8)|(2n+3)$ by assumption, we use Exercise 2B (with $k=3$ and $l=-2$) to see that $(3n-8)$ divides $3(2n+3)-2(3n-8)=25$. If $3n-8$ divides 25, then $3n-8$ must equal ± 1, ± 5, or ± 25. Since $\frac{2n+3}{3n-8}$ is a positive integer, there are only 3 choices available:

$$3n-8 = +1, +5, \text{ or } +25$$

so $3n$ must be 9, 13, or 33 or $n=3$, $\frac{13}{3}$ or 11.

Because n is a positive integer by assumption, the only possibilities left are $n=3$ or 11 and they both work.   //

Here is an alternative solution of Example 8 which doesn't use Exercise 2B. Let's define m as

$$\frac{2n+3}{3n-8} = m \in \mathbb{N}.$$

Then $2n+3 = m(3n-8)$ or $3 = 3mn - 2n - 8m$.

To aid in the factoring process, we multiply the last equation by 3 and adding 16 to both sides gives

$$25 = 9mn - 6n - 24m + 16.$$

The right-hand side of the previous equation factors into $(3n-8)(3m-2)$. Thus $(3n-8)(3m-2)$ must have the same divisors as 25. The only choices for solutions of integer values are summarized in the table which follows:

$3n-8$	1	-1	5	-5	25	-25
$3m-2$	25	-25	5	-5	1	-1

(This means, for example, that $3n-8=1$, and $3m-2=25$ is one possibility to check out). Examining each possibility gives:

$3n$	9	7	13	3	33	-17
$3m$	27	-23	7	-3	3	1

Since both n and m are assumed to be positive integers, there are only two solutions remaining

$$(n=3 \text{ and } m=9) \text{ or } (n=11 \text{ and } m=1).$$

In other words, $n=3$ or 11 are only solutions (as before).

Note: Example 8 can also be solved in the following simple way. Since $\frac{2n+3}{3n-8} \in \mathbb{N}$ then $\frac{6n+9}{3n-8} = 2 + \frac{25}{3n-8} \in \mathbb{N}$. Then $\frac{25}{3n-8} \in \mathbb{N}$, hence $3n-8 = 1, 5$, or 25 which gives $n = 3, 11$, or $\frac{13}{3}$. Thus $n = 3$ or 11 are the only solutions in $\mathbb{N}$.

Exercise 7. Let n be a non-negative integer. Find all n such that $(2n-1)|(6n+5)$.

The purpose of looking at this example with two approaches is to show the ways in which a result like Exercise 2B can simplify proofs. While the second method that was used to solve Example 8 is elementary, it required some "tricks" to complete. Having proven Exercise 2B gives a structure for looking at this example and a platform for proving other results. Problem solving can be "tricks" or can use mathematical structure—both are correct. It becomes a choice as to which you prefer.

One should admire the cleverness of the proof or solution of some results. We believe, however, that a major aspect of the appeal and beauty of mathematics is in its coherence and the way it builds on its foundation. Of course, what each of us sees as the beauty of mathematics is very personal.

Problem Set 1.2:

1. (a) Let $n \in \mathbb{N}$ be odd. Prove that the sum of any n consecutive numbers is divisible by n (Hint: Let x be the first number and write out the sum).
   (b) Is the result of Part (a) true for any even integer of $n \geq 3$?
2. Find all integers x such that $(2x-1)|(9x-4)$.
3. Find all $x \in \mathbb{Z}$ such that $(3x+1)|(5x-2)$.
4. Assume that n is an integer greater than 1. For what values n is $\frac{2n+1}{n-1}$ an integer?
5. Find all positive integers a and b such that $ab = a - 5b + 20$. (Hint: What is $(a+5)(b-1)$?)
6. (a) Find all positive integers a and b such that $ab - a - 3b = 4$.
   (b) Find all positive integers a and b such that $a(b+2) = b$.
7. (a) Let $m \in \mathbb{Z}$. If $m|(21n+20)$ and $m|(7n+3)$ for some $n \in \mathbb{Z}$, what are the possible value(s) of m?
   (b) Let $b \in \mathbb{Z}$ and assume $b|(b+8)$, $(b-1)|(b+11)$ and $(b-4)|(3b+6)$. Find the value of b.
8. Using Exercise 2B, find all positive integers n such that $\frac{4n-1}{2n+3}$ is a positive integer.
9. Using the method of the alternate solution from Example 8 and the simple way described in the note following Example 8, find all positive integers n such that $\frac{4n-1}{2n+3}$ is a positive integer.
10. Using the method of the alternate solution from Example 8 and the simple way described in the note following Example 8, find all positive integers n such that $\frac{3n+1}{2n-6}$ is a positive integer.
11. Use all methods to find all possible positive integers n such that $\frac{2n+8}{n-7}$ is a positive integer.

12. Use all methods (similar to those above) to find all positive integers n such that

$$\frac{2n}{3n - 5} \text{ is a positive integer.}$$

13. Let a, b, c, d, e and k be any integers. If e divides $(ka - b)$ and e divides $(kc - d)$, prove that e divides $(ad - bc)$.

1.3 Long Division: The Division Algorithm

Overview Our next topic is "long division". It is an idea which you will meet again and again during mathematics and possibly a teaching career in many different ways. The procedural understanding of how to actually do long division is, of course, very important. Conceptual understanding of this process is also crucial.

Many parts of problem solving in number theory are based on an in-depth comprehension of long division. We will work with both an example and a statement using the variables a and b with the quotient being q and the remainder r in order to show conceptually what long division is. For example, we know that 37 divided by 7 has a quotient $q = 5$ and remainder $r = 2$.

If a and b are integers, with $a > b$ and b positive, doing what is usually called "long division" is a way to check if $b|a$. The foundation for long division is to start with the integer b, subtract it from a, and then repeat the process. To illustrate the procedure, we will use $b = 7$ and $a = 37$ as an example. We keep subtracting the number b from a and then take b from what remains, again and again. How many times? The answer is q times ($q \in \mathbb{N}$) until either $a = qb$ (which isn't the case for $a = 37$ and $b = 7$) or $a = qb + r$ where $0 < r < b$ ($q = 5$ and $r = 2$ in the case of $b = 7$ and $a = 37$). This is formalized by the Division Algorithm, which is so important that we don't give it a number, just a name, and mark it with a star ($*$).

The Division Algorithm. *Let* $a \in \mathbb{Z}$, $b \in \mathbb{N}$, *then there exist integers* q *and* r *such that*

$$a = qb + r \text{ and } 0 \leq r < b. \tag{$*$}$$

Furthermore, the integers q *and* r *which solve Eq. ($*$) are uniquely determined.*

Definition. The integer q in Division Algorithm ($*$) is called the QUOTIENT of a by b and r is called the REMAINDER of a on division by b. Note that $b|a$ exactly when the remainder of a by b is zero. We say a is the DIVIDEND and b is the DIVISOR.

From the sketch of how the Division Algorithm works in the case of $b = 7$ and $a = 37$ which precedes its statement, an argument can be made (if $a > 0$) that it should be called the "successive subtraction" algorithm, instead of the Division

Algorithm. The first three problems of Problem Set 1.3 show the "successive subtraction" approach for $a > 0$.

Suppose we have two integers n_1 and n_2. We will now look into what the remainder of their sum and the remainder of their product on division by b are as that will be useful in what follows. Simple algebra shows that if q_1, q_2, r_1, and r_2 are integers and (using the conventions of the Division Algorithm) both:

$$n_1 = q_1 b + r_1 \text{ with } 0 \le r_1 < b$$

and

$$n_2 = q_2 b + r_2 \text{ with } 0 \le r_2 < b,$$

then

$$n_1 + n_2 = (q_1 + q_2)b + (r_1 + r_2) \tag{1.3.1}$$

and

$$n_1 n_2 = q_1 q_2 b^2 + (q_1 r_2 + q_2 r_1)b + r_1 r_2. \tag{1.3.2}$$

The first term of the right side of (1.3.1) is divisible by b and the first two terms of the right side of (1.3.2) are divisible by b. Therefore we have the useful result contained in the next proposition.

Proposition 2. *If n_1 and n_2 are positive integers whose remainders on division by the positive integer b are r_1 and r_2 respectively, then*

$$(n_1 + n_2) \text{ has the same remainder on division by } b \text{ as } (r_1 + r_2) \text{ has} \tag{1.3.3}$$

and

$$n_1 n_2 \text{ has the same remainder on division by } b \text{ as } r_1 r_2 \text{ has.} \tag{1.3.4}$$

Proof. The Division Algorithm shows that there are unique integers q_1, q_2, r_1, and r_2 such that

$$n_i = q_i b + r_i \text{ with } 0 \le r_i < b \text{ for } i = 1, 2. \tag{1.3.5}$$

We first prove Eq. (1.3.3) by adding the equations of (1.3.5) to obtain

$$n_1 + n_2 = (q_1 + q_2)b + (r_1 + r_2). \tag{1.3.6}$$

However, we can apply the Division Algorithm to the integer $(r_1 + r_1)$. Writing the quotient or $r_1 + r_2$ on division by b as q^* and the

$$\text{remainder of } (r_1 + r_2) \text{ on division by } b = r^* \qquad (1.3.7)$$

and

$$r_1 + r_2 = q^* b + r^* \text{ with } 0 \leq r^* < b.$$

Substituting into Eq. (1.3.6) yields

$$(n_1 + n_2) = (q_1 + q_2 + q^*) b + r^*. \qquad (1.3.8)$$

Equation (1.3.8) writes the integer $(n_1 + n_2)$ as the sum of an integer times b plus an integer r^* with $0 \leq r^* < b$. The quotient and remainder of any integer, in particular, $(n_1 + n_2)$ on division by b *must be unique* so Eq. (1.3.8) shows that the quotient of $(n_1 + n_2)$ is $(q_1 + q_2 + q^*)$ and the remainder of $(n_1 + n_2)$ on division by b is r^*, which by definition is Eq. (1.3.7), the remainder of $(r_1 + r_2)$ on division by b.

To prove Eq. (1.3.4) we follow the same procedure.

$$n_1 n_2 = q_1 q_2 b^2 + (q_1 r_1 + q_2 r_2) b + r_1 r_2. \qquad (1.3.9)$$

However, if $r =$ remainder of $r_1 r_2$ on division by b and q is its quotient, then

$$r_1 r_2 = q b + r \text{ where } 0 \leq r < b.$$

Substituting into Eq. (1.3.9) yields

$$n_1 n_2 = (q_1 q_2 b + q_1 r_2 + q_2 r_1 + q) b + r \text{ with } 0 \leq r < b.$$

Thus, by uniqueness, r must be the remainder of $n_1 n_2$ on division by b. Of course, $q_1 q_2 b + q_1 r_2 + q_2 r_1 + q$ is the quotient of $n_1 n_2$ by b, but that result is not useful. ∎

Exercise 8.

A. Explain why in Eq. (1.3.3) we did not claim that the remainder of $n_1 + n_2$ is always equal to $r_1 + r_2$. (Give an example.) Hint: Look at the statement of the Division Algorithm.
B. Explain why in Eq. (1.3.4) we did not claim that the remainder of $(n_1 n_2)$ is always $r_1 r_2$. (Give an example.)

Example 9. Find the remainder of

$$l = 1841 + 1954$$

on division by 6.

Solution. Since $1841 = (306)6 + 5$ and $1954 = (325)6 + 4$, the remainder of $l = 3,795$ on division by 6 therefore is the same as the remainder of $(5 + 4) = 9$ by 6. The answer is 3; that is, the remainder of 3795 on division by 6 is 3. This problem gives an example of why the remainder of $(n_1 + n_2)$ is not $r_1 + r_2$.   //

Example 10. Find the remainder of $n = (1841)(1954)$ on division by 5.

Solution. Since $1841 = 1840 + 1$, the remainder of 1841 on division by 5 is 1. The remainder of $1954 = 1950 + 4$ is 4, so Proposition 2 says that the remainder of their product n, is the product of $(4)(1) = 4$ (since $4 < 5$).   //

Example 11. Find (a) the remainder of $n = (40)^5$ on division by 19; (b) the remainder of $n = (40)^{10}$ on division by 19.

Solution.

(a) The remainder of 40 on division by 19 is 2. Repeated application of Proposition 2 shows that remainder of $(40)^5$ on division by 19 is the same as the remainder of $2^5 = 32$ on division by 19 and so the answer is 13.
(b) Since the remainder of 40 on division by 19 is 2, the remainder of $n = (40)^{10}$ on division by 19 is the same as the remainder of $2^{10} = (2^5)^2$ by 19. Because we know the remainder of 2^5 by 19 is 13 from part (a), we need only find the remainder of $(13)^2 = 169$ by 19. Since $169 = (19)(8) + 17$, the answer is 17. This problem gives an example to show why the remainder of $(n_1 n_2)$ need not be $r_1 r_2$.

                                                                    //

In the Division Algorithm, we have assumed that the divisor b is positive but here we made no assumption about the dividend a. What then is the interpretation of the Division Algorithm if $a < 0$? Let's start with an example.

Example 12. What is the quotient and remainder of -5 on division by 2?

Solution. The precise conclusion of the Division Algorithm means that we must find $q \in \mathbb{Z}$ and $r \in \mathbb{N} \cup \{0\}$ with

$$-5 = q2 + r \qquad (1.3.10)$$

where $0 \leq r < 2$. In the interpretation of the case in which a and b are both positive, we used subtraction of b from a. However, subtraction of 2 from -5 gives -7. We are going in the wrong direction to get the quotient and remainder of -5 on division by 2. Instead, addition will help:

$$-5 + b = -3 \text{ but } -3 \text{ is not between 0 and 2} \qquad (1.3.11)$$

$$-3 + b = -1 \text{ but } -1 \text{ is not between 0 and 2} \qquad (1.3.12)$$

$$-1 + b = 1 = r \text{ which satisfies } 0 \leq r < 2. \qquad (1.3.13)$$

How many additions were there? There were 3 so we are led to $q = -3$ and $r = 1$.
Putting these values for q and r into Eq. (1.3.10) gives

$$-5 = (-3)2 + 1$$

which checks. Thus the remainder is 1 and the quotient is -3. Note that there is no
assumption that the quotient must be positive in the Division Algorithm.   //

Problem Set 1.3:

1. (a) What are the quotient and remainder of 63 divided by 17?
   (b) Follow the discussion at the beginning of this section and first subtract 17
       from 63, then subtract 17 from your answer, etc. until you end up with
       an integer which is at least 0 but less than 17. Show that the number of
       subtractions is the quotient of 63 by 17 and the remainder is the number
       between 0 and 17.
2. (a) What are the quotient and remainder of 35 by 7?
   (b) Repeat part b of Problem 1 above.
3. (a) With the help of Example 12, what are the quotient and remainder of -27
       by 6?
   (b) Repeat part b of Problem 1 above.
4. (a) What are the quotient and remainder of -7 divided by 3?
   (b) What happens if you try to interpret the quotient of -7 by 3 using
       successive subtractions?
   (c) What happens if you try to interpret the quotient of -7 by 3 using
       successive additions?
5. (a) What are the quotient and remainder of -5 divided by 3?
   (b) What happens if you try to interpret the quotient of -5 by 3 using
       successive subtractions?
   (c) What happens if you try to interpret the quotient of -5 by 3 using
       successive additions?
6. (a) What are the quotient and remainder of -32 divided by 6?
   (b) What happens if you try to interpret the quotient of -32 by using successive
       subtractions?
   (c) What happens if you try to interpret the quotient of -32 by 6 using
       successive additions?
7. Without using a calculator,

   (a) find the remainder of $(12345)(678)$ on division by 13.
   (b) find the remainder of $(30)^9$ on division by 13.

8. Find the remainder of $n = (12345)(678) + 30^9$ on division by 13.
9. Without using a calculator,

   (a) find the remainder of $(15743)(365)$ on division by 7.
   (b) find the remainder of $(50)^{15}$ on division by 49.

10. Find the remainder of $n = (15743)(365) + 50^{15}$ on division by 7.
11. Find the remainder of $n = (1841)(1954) + 40^{10}$ on division by 11.
12. Suppose that y is the sum of squares of *three* consecutive integers. Prove that the remainder of y on division by *three* is 2.
13. Suppose that y is the sum of squares of any *four* consecutive integers. Prove that the remainder of y on division by *four* is 2.
14. (a) Given the results of Problems 12 and 13, do you think it is reasonable to make the following conclusion?

    Conclusion Let y be the sum of squares of n consecutive integers ($n > 2$). Then the remainder of y on division by n is 2.
    (The answer to this question is yes or no! That is, yes, it sounds reasonable or no, it doesn't sound reasonable.)

    (b) Is the above conjecture true? If so then prove it, otherwise find a counterexample. (Hint: Try $n = 5$.)
    Fact[1]: There is a formula due to Jakob Bernoulli (1654–1705) that gives the sum of squares. If $n \geq 2$ is an integer then

    $$1^2 + 2^2 + \cdots + (n-1)^2 = \frac{n(n-1)(2n-1)}{6}.$$

15. Suppose y is the sum of the squares of n consecutive integers $n \geq 2$ and x is the first integer.

    (a) Show that

    $$y = nx^2 + n(n-1)x + \frac{n(n-1)(2n-1)}{6}.$$

    Hint: Use Bernoulli's formula of Problem 14 for the sum of squares above.
    (b) Show that the remainder of y on division by n is the remainder of $r = \frac{n(n-1)(2n-1)}{6}$ on division by n (for example, $n = 4$ gives $r = 14$, so the remainder of y on division by n is 2).
    (c) Can you conclude from the formula (a) that y is divisible by n? Why or why not?

1.4   Tests for Divisibility in Base Ten

Overview Here's a general class of problems to solve: how can we determine whether a given (small) integer can divide the integer n? This is the "problem of divisibility". This subject is another example in which problem solving depends on

---

[1]This fact can be proven easily if the reader is familiar with mathematical induction.

establishing an approach (a foundation) from which we can directly solve problems. This section makes some definitions and then solves the problem of divisibility by all integers up to 13.

Divisibility Problem Overview How can we determine when an integer k divides another integer n? What properties of n guarantee that k divides n?

The answer appears by looking at combinations of digits in our base ten representation of n. This section uses the base ten representation of integers to come up with criteria for divisibility by many small integers.

For what k can we come up with a test for whether k divides n? The criteria of the tests for divisibility will involve the sum and difference of the digits that form the base ten representation. In the rest of this section, we will present tests for: $k = 2, 3, 4, 5, 7, 8, 9, 11, 13$ and all powers of 2 and 5. The list is more complete than it looks because for $k = 1$, there is nothing needed since for any n, k is a divisor! Furthermore, $6 = 2 \cdot 3$ and so we can use the test for $k = 2$ and $k = 3$ to investigate divisibility by 6. Since an integer is divisible by ten exactly when its last digit is zero, we will have test for divisibility by all integers from 2 to 13 as well as some others. See the problems at the end of this section. We will see in other examples (for instance, Example 17) that to prove divisibility by a composite, it suffices to check divisibility by its prime divisors.

Definition. The collection of integers $a_0, a_1, \ldots, a_l$ each of which is between zero and nine is called the BASE TEN REPRESENTATION of n if Eq. (1.4.1) is satisfied. The integer n is written informally as $n = a_l a_{l-1} \cdots a_2 a_1 a_0$ and $a_l, \ldots, a_0$ are called DIGITS. In particular, a_0 is called the LAST DIGIT of n. More formally,

$$n = a_l 10^l + a_{l-1} 10^{l-1} + \cdots + a_2 10^2 + a_1 10^1 + a_0 \text{ and } a_l \neq 0. \qquad (1.4.1)$$

The tests below are named by the integer that we hope will divide the given number. Except for Test 2, each test is illustrated by an example before its proof is given. At that time, we restate each test to fix the notation of the proof.

Test 2 A positive integer n is divisible by 2 if and only if the last digit of n is even. In general, $n \in \mathbb{N}$ is divisible by 2^k if and only if the last k digits of n are divisible by 2^k. Equivalently, let the base 10 representation of n be $n = a_l a_{l-1} \ldots a_0$. Then $2^k | n$ if and only if $2^k | a_{k-1} \ldots a_0$.

Exercise 9. Use Test 2 to decide if 1336 is divisible by 2. Is it divisible by 4? Is it divisible by 8?

Test 5 A positive integer is divisible by 5 if and only if the last digit is divisible by 5 (i.e. the last digit is either 0 or 5). In general, $n \in \mathbb{N}$ is divisible by 5^k if and only if the last k digits of n are divisible by 5^k.

Example 13. Show that 375 is divisible by 5, 25, and 125.

Solution. Since $5 = 5^1$, $25 = 5^2$, and $125 = 5^3$, we need to determine, thanks to Test 5, whether $5|5$, $5^2|75$ and $5^3|375$, all of which are true. Thus 375 is divisible by 5, 25 and 125.   //

Restatement of Test 5 If the base 10 representation of n is $n = a_\ell a_{\ell-1} \cdots a_2 a_1 a_0$ then $5^k | n$ if and only if $5^k | a_{k-1} a_{k-2} \cdots a_0$.

Proof. We will leave the case when $k = 1$ and $k = 3$ as Exercise 10 and work through the case when $k = 2$. The general case can easily be proven using mathematical induction if the reader has studied that concept. When $k = 2$, the theorem (assuming $\ell \geq 2$) says that $25|n$ if and only if 25 divides the integer represented by $a_1 a_0$ which is $10a_1 + a_0$. From (1.4.1)

$$n = \left(a_\ell 10^\ell + a_{\ell-1} 10^{\ell-1} + \cdots + a_2 10^2\right) + (10a_1 + a_0).$$

But every term in the first parenthesis has a factor of 100. Thus $25|n$ if and only if $25|(10a_1 + a_0)$. Since $10a_1 + a_0$ has as its base 10 representation $a_1 a_0$, we are finished. ∎

Exercise 10. If the base 10 representation of n is $n = a_\ell a_{\ell-1} \cdots a_2 a_1 a_0$, prove that

A. $5|n$ if and only if $5|a_0$. (Test 5 for $k = 1$)
B. $125|n$ if and only if 125 divides the integer whose base ten representation if $a_2 a_1 a_0$. (Test 5 for $k = 3$)

The next three tests involve looking at the sums (or differences) of the digits that represent the given number, n, in base 10. Before starting these tests, we will need some terminology. There is a slight technicality depending on whether there is an even number or an odd number of digits in the base ten representation of n.

Definition. If $n \in N$ is represented as a base 10 number as in Eq. (1.4.1) via

$$n = a_l a_{l-1} \cdots a_2 a_1 a_0$$

then the DIGIT SUM OF n is $a_l + a_{l-1} + \cdots + a_2 + a_1 + a_0$. If l is even, then

the EVEN DIGIT SUM is $a_0 + a_2 + a_4 + \cdots + a_{\ell-2} + a_\ell$

and

the ODD DIGIT SUM is $a_1 + a_3 \cdots + a_{\ell-3} + a_{\ell-1}$.

If l is odd, the even digit sum and odd digit sum are defined analogously (but end with a_{l-1} and a_l respectively). The even digit sum always includes a_0 and the other even-indexed digits, whereas the odd digit sum includes a_1 and the other odd-indexed digits.

Example 14. If $n = 3574$ then there are 4 digits in the base 10 representation of n so that $\ell = 3$. The digit sum is 19, the odd digit sum is 10, and the even digit sum is 9. If $m = 3, 556, 143$ then $l = 6$ and the even digit sum is $3 + 5 + 1 + 3 = 12$ and the odd digit sum is $5 + 6 + 4 = 15$.

Test 3/Test 9 (Test for divisibility by 3 or 9): A positive integer n is divisible by 3 if and only if its digit sum is divisible by 3. The integer n is divisible by 9 if and only if its digit sum is divisible by 9.

Example 14 (continued) Since the digit sum of $n = 3574$ is 19 and 19 is not divisible by 3, 3 does not divide 3574. Since the digit sum for $m = 3,556,143$ is 27, m is divisible by 3 and 9.

Restatement of Test 3/9 Let n be represented as $n = a_\ell a_{\ell-1} \cdots a_2 a_1 a_0$. Then we have the following two results:

A. 3 divides n if and only if 3 divides $(a_\ell + a_{\ell-1} + \cdots + a_2 + a_1 + a_0)$
B. 9 divides n if and only if 9 divides $(a_\ell + a_{\ell-1} + \cdots + a_2 + a_1 + a_0)$.

Proof. Part B will be left as an exercise (Problem 6 of Problem Set 1.4) since its proof is very similar to Part A. We'll now show that the remainder of n on division by 3 is the same as the remainder of the digit sum of n on division by 3. Note that the remainder of 10 on division by 3 is 1. Hence, by Proposition 2, the remainder of 10^2, 10^3, etc. is also 1. Since n is given by Eq. (1.4.1) invoking both Eqs. (1.3.3) and (1.3.4) shows that

$$\text{remainder of } n \text{ on division by } 3 = (\text{remainder of } 10^\ell)(\text{remainder of } a_\ell)+$$

$$(\text{remainder of } 10^{\ell-1})(\text{remainder of } a_{\ell-1}) + \cdots$$

$$+(\text{remainder of } 10^2)(\text{remainder of } a_2)$$

$$+(\text{remainder of } 10)(\text{remainder of } a_1) + \text{remainder of } a_0.$$
$$(1.4.2)$$

But the first term of each product of Eq. (1.4.2) is one from the first paragraph of the proof so the remainder of n on division by 3 is the remainder of $a_\ell + a_{\ell-1} + \cdots + a_0$ on division by 3. ∎

Included in the proof is the stronger result which we give as Corollary 2.

Corollary 2.

A. *The remainder of n on division by 3 is the same as the remainder of the digit sum of n on division by 3.*
B. *The remainder of n on division by 9 is the same as the remainder of the digit sum of n on division by 9.*

Using Corollary 2B is also called the method of "casting out nines". Many grade school children are fascinated by this test.

Exercise 11. Find all values of x in each of the following.

A. $3|435x17$.
B. $9|435x7$.

Test 11 A positive integer n is divisible by 11 if and only if the difference between its even digit sum and odd digit sum is divisible by 11.

Example 15. Let $n = 6853$. Then the even digit sum is $3 + 8 = 11$, the odd digit sum is $5 + 6 = 11$ and the difference is $11 - 11 = 0$. Because 11 divides 0, 11 divides 6853.

Remembering that zero is divisible by 11 (since $11 \cdot 0 = 0$), do the following exercise. Of course, zero is divisible by all integers except 0.

Exercise 12. Which of the following integers are divisible by 11?

A. 2,343
B. 9,190,718,290
C. $n = 34c67c41$

We'll restate Test 11 in the case where ℓ is even. When ℓ is odd, the proof is the same—it is just the notation that is slightly different.

Restatement of Test 11 If the base 10 representation of n is $n = a_\ell a_{\ell-1} \cdots a_2 a_1 a_0$ then $11|n$ if and only if

$$11|((a_\ell + a_{\ell-2} + \cdots + a_2 + a_0) - (a_{\ell-1} + a_{\ell-3} + \cdots a_3 + a_1)).$$

Proof. We'll use the simple observation that $10 = 11 - 1$ with the obvious fact that we can rewrite Eq. (1.4.1) as

$$n = a_l(11 - 1)^l + a_{l-1}(11 - 1)^{l-1} + \ldots + a_1(11 - 1) + a_0. \tag{1.4.3}$$

Note that any power of $(11 - 1)$ has 11 in each term except the last one which is ± 1 (expand the term $(11 - 1)^k$ using algebra without simplifying). More precisely, there are integers $b_k \in \mathbb{Z}$ with

$$(11 - 1)^k = 11(\text{some integer}) \pm 1 = 11b_k \pm 1$$

where $+1$ is used if k is even and -1 is used if k is odd. When l is even, for example, Eq. (1.4.3) becomes

$$n = a_l(11b_l - 1)^l + a_{l-1}(11b_{l-1} - 1)^{l-1} + \ldots + a_1(11b_1 - 1) + a_0 \tag{1.4.4}$$

or

$$n = 11(a_l b_l + a_{l-1}b_{l-1} + \ldots + a_1 b_1)$$
$$+(a_l - a_{l-1} + \ldots + a_2 - a_1 + a_0). \tag{1.4.5}$$

Thus $11|n$ if and only if 11 divides the alternating sum of the base 10 representation of n which is Test 11. The case in which l is odd is exactly the same. ∎

Because it is used in practice, we isolate the definition implicit in this proof.

Definition. The ALTERNATING DIGIT SUM OF n is the difference of the even digit sum and the odd digit sum.

Definition. Suppose that $n = a_l 10^l + a_{l-1} 10^{l-1} + \cdots + a_0$ so that n is written as $a_l a_{l-1} \cdots a_1 a_0$ in base ten representation. Consider the ASSOCIATED SEQUENTIAL TRIPLE SUM for n, $s^T(n)$, defined as the finite sum:

$$s^T(n) = (a_0 + 10a_1 + 10^2 a_2) - (a_3 + 10a_4 + 10^2 a_5)$$
$$+ (a_6 + 10a_7 + 10^2 a_8) - (a_9 + 10a_{10} + 10^2 a_{11}) + \cdots$$

Let's look at Example 16A more closely to see a pattern which makes it much easier to follow the definition of the sequential triple sum. In that case when $n = 176, 182, 461$, writing vertically,

$$s^T(n) = 461$$
$$-182$$
$$+\underline{176}$$
$$455$$

In other words, starting with the integer n, we look at the 3 digits between each of the commas and form their alternating sum. This observation will be useful to you as you work homework problems.

Test 7/Test 13 n is divisible by:

(a) 7 if and only if $s^T(n)$ is divisible by 7.
(b) 13 if and only if $s^T(n)$ is divisible by 13.

Example 16.

A. Determine whether $n = 176, 182, 461$ is divisible by 7, without the use of a calculator.
B. For what values of a is

$$n = 14, 4a6, 3a2, 121$$

divisible by 13?

Solution.

A. The associated sequential triple sum for n is

$$s^T(n) = (1 + 6(10) + 4(100))$$
$$- (2 + 8(10) + 1(100)) + (6 + 7(10) + 1(100))$$
$$= 461 - (182) + (176) = 455.$$

Since $7 | 455$, the original n is divisible by 7. (The power of this method is that we've reduced the question of divisibility of a 9 digit number to that of a 3 digit number.)

B. Once again, forming the associated sequential triple sum gives

$$s^T(n) = 121 - (3a2) + (4a6) - 14$$

$$= 107 - (300 + 10a + 2) + (400 + 10a + 6)$$

$$= 211.$$

Since 211 is not divisible by 13, there are no values of a for which the given n of Part B is divisible by 13.

$$//$$

Since the notion can become cumbersome (to say the least!) we'll now prove Test 7/13 only for the case of a 6 digit number (so $\ell = 5$). What is astounding is that both of these tests can be proven simultaneously because of the clever observation that

$$1001 = 7(11)(13).$$

Restatement of Test 7/13   Suppose that n is a six digit number whose base ten representation is $n = a_5a_4a_3a_2a_1a_0$ and write

$$s^T(n) = (a_0 + 10a_1 + 10^2a_2) - (a_3 + 10a_4 + 10^2a_5). \qquad (1.4.6)$$

Show that:

(a) n is divisible by 7 if and only if $s^T(n)$ is divisible by 7
(b) n is divisible by 13 if and only if $s^T(n)$ is divisible by 13.

Proof. We will use the observation that

$$10^3 = (1001 - 1) = (7 \cdot 11 \cdot 13) - 1. \qquad (1.4.7)$$

We'll start with $n = a_5a_4a_3a_2a_1a_0$ so that

$$n = (a_0 + 10a_1 + 10^2a_2) + (10^3a_3 + 10^4a_4 + 10^5a_5)$$

$$= (a_0 + 10a_1 + 10^2a_2) + 10^3(a_3 + 10a_4 + 10^2a_5)$$

$$= (a_0 + 10a_1 + 10^2a_2) + (1001 - 1)(a_3 + 10a_4 + 10^2a_5)$$

$$= (a_0 + 10a_1 + 10^2a_2) + (7 \cdot 11 \cdot 13)(a_3 + 10a_4 + 10^2a_5)$$

$$- (a_3 + 10a_4 + 10^2a_5)$$

where the last equality follows from Eq. (1.4.7). Thus, from the definition of the sequential triple sum and Eq. (1.4.7), we get

$$n = s^T(n) + (7 \cdot 13 \cdot 11)(a_3 + 10a_4 + 10^2 a_5).$$

The second term of the right hand side is always divisible by both 7 and 13 (and 11) so n is divisible by 7 (or 13) if and only if $s^T(n)$ is divisible by 7 (or 13). In fact, we're showing that the remainder of n on division by 7 (or 13) is the same as the remainder of $s^T(n)$ on division by 7 (or 13). A truly remarkable proof!   ∎

It's important to note two points. First of all, the associated sequential triple sum really reduces the question of divisibility of the given number by 7 (no matter how many digits it has) to the divisibility of a much smaller (frequently three digit) number by 7—a much more manageable question.

Secondly, without the theory developed in the above tests, either part of Example 16 would be very hard. It is the foundation which we have developed (Test 7/13) which allows for easy problem solving. Once again, an established theory is a tremendous help in solving problems and can reduce the problem to an easy exercise.

Exercise 13. Determine whether the integer of Exercise 12B is divisible by 13.

To do a problem involving divisibility by a composite number, one may check divisibility by divisors that are prime numbers (and their powers if needed). That is what the Fundamental Theorem of Arithmetic (Theorem 1) enables us to do. Recall that we used this major theorem in the case when we divided by $6 = 2 \cdot 3$.

Example 17. Let n be represented in base 10 as $12a3b$ so that $n = 10^4 + 2(10^3) + a(10^2) + 3(10) + b$. Find all digits a and b such that $15|n$ and find n.

Solution. Note that $15|n$ if and only if $3|n$ and $5|n$. We will use Tests 3 and 5 on the integer $n = 12a3b$ whose digit sum is $s(n) = 6 + a + b$. If $5|12a3b$ then Test 5 says there are only two possibilities, $b = 0$ or $b = 5$.

If $b = 0$, then $s(n) = 6 + a$. Thus, when $b = 0$, $3|n$ if and only if $3|(6 + a)$ which simplifies to $3|a$. Thus $a = 0,3,6,9$. Therefore, the solutions for $15|n$ when $b = 0$ are

$$n = 12030;\ 12330;\ 12630;\ \text{or } 12930.$$

If $b = 5$, a similar argument shows that $a = 1,4,7$. Put differently, the solutions for n (with $b = 5$) are: $n = 12135;\ 12435;\ \text{or } 12735.$

All the possible values for a and b and the corresponding number n are summarized in the table below:

a	0	3	6	9	1	4	7
b	0	0	0	0	5	5	5
n	12,030	12,330	12,630	12,930	12,135	12,435	12,735

//

Exercise 14. If 3 divides $n = 34a56b$ and 4 divides n find the digits a and b. If there is more that one solution, find them all. If there are none, explain why.

Example 18. If $44 | 1985c63d$ find all possible values for the digits c and d.

Solution. Now $44 = 4 \cdot 11$ and so we'll start with 4 (the "easier factor"). Since $2^2 = 4$ divides the number, its last two digits must be 32 or 36 so either $d = 2$ or $d = 6$.

We now use Test 11 and $n = 1985c63d$. We see $11 | n$ exactly when 11 divides $((1 + 8 + c + 3) - (9 + 5 + 6 + d))$ or $11 | (c - d - 8)$. If $d = 2$ then $11 | (c - 10)$ which is impossible if c is to be a digit (since $c \leq 9$).

The case in which $d = 6$ yields a solution. When $c = 3$ (and for no other digits) $11 | (c - 14)$ so the only solution is $c = 3$ and $d = 6$ (which means that $n = 19,853,636$. $\qquad\qquad //$

A word of caution about the tests given in this section. They depend critically on the representation of the integer n being base-10. For example, the number 4 is divisible by 2 and so its base-3 representation must also be divisible by 2. If you have already seen base-3 arithmetic, because, $4 = 1 \cdot 3 + 1$, so the base-3 representation of 4 is 11. The last digit of the base-3 representation of 4 is not divisible by 2. This example does not contradict Test 2 because Test 2 only deals with base-ten representations. We now know that

$$11 \text{ (base-3)} = 2 \text{ (base-3) times } 2 \text{ (base-3)}$$

or

$$11_3 = 2_3 \text{ times } 2_3.$$

and so 2 does divide 11 (base-3). Whether a number (2 in this case) divides another (4 in this case) is independent of what the base of that number is. The test may change depending on the base whereas whether a number is divisible by another does not change.

In the remainder of this chapter, we will discuss the divisibility tests when the number given in a base other than 10.

Problem Set 1.4:

1. If n is written as $12a4c$ in base 10, find all digits a and c such that 15 divides n.
2. If $33 | 1a24b$, find all possible digits a and b.
3. If $44 | 1985ab3b$, find all possible digits a and b.
4. If n is the 6 digit number $n = a2b241$ and $117 | n$, find all possible values for n. (Hint: $117 = 13 \cdot 9$.)
5. Prove Test 10: the integer n is divisible by 10 if and only if its last digit is zero.
6. Prove Test 9.
7. If $45 | 41a45b$, find all possible digits a and b.

8. If $36|12a49b$, find all possible digits a and b.

9. If 48 divides $n = 18a5b6$ and 11 divides n, find all possible digits a and b. If there are none, explain why.

10. If 98 divides $m = 62c11d$ and 3 divides m, find all possible digits c and d. If there are none, explain why.

11. If 63 divides $n = 2266ab$ and 4 divides n, find all possible digits a and b. If there are none, explain why.

12. If $12a3b6$ is divisible by 21 and $b > a$, find all possible digits a and b and show there are no more.

13. Let n be given and m be the integer whose base 10 representation reverses the digits of n. Prove or disprove each of the following.

    (a) If $a|n$ then $a|m$.
    (b) If $11|n$ then $11|m$.
    (c) If $2^5|n$ then $2^5|m$.
    (d) If $7|n$ then $7|m$

14. If $n \in \mathbb{N}$, show that 6 divides $7^n - 1$. Can you generalize this result?

15. If p and q are primes and $p \neq q$, write a test for divisibility by pq.

16. Find all "palindromic" 5-digit numbers which start with 2 and are divisible by 18. (A number is a palindrome if it reads the same backwards as forwards; that is, if its first and last digits are the same, second and next to last digits are the same, and so on. For instance, 464 and 627726 are palindromes.)

17. Prove that any palindrome with an even number of digits is divisible by 11.

18. Let $abcd$ be a 4-digit positive integer. Prove that $11|(abcd + bcda)$.

19. Without using a calculator, determine which of the following numbers is a multiple of 9.(a) 247023846 (b) 645×7329 (c) $986^3 + 814^3$

20. Find all choices for the digits a and c so that n, which is a five digit number represented by $a56c2$, is divisible by 9.

1.5 Binary and Other Number Systems

Overview This section is not necessary nor dependent on other sections, but it is quite interesting. In this section, we look at arithmetic in a base b which is not the usual $b = 10$ basis. Examining numbers in a base other than ten gives depth to the understanding of the basic concepts of arithmetic and expands our minds. We will finish this section by looking at what happens to the divisibility tests in base ten of Sect. 1.4 when a base not equal to ten is used.

The most common and simplest representation of integers is the DECIMAL (or BASE TEN) system, using the ten digit symbols 0, 1, 2, 3, 4, 5, 6, 7, 8, and 9. In this system, the digits of an integer represent multiples of powers of 10 (including $10^0 = 1$). For example:

$$2005 = 2 \cdot 10^3 + 0 \cdot 10^2 + 0 \cdot 10^1 + 5 \cdot 10^0 \ (10^0 = 1)$$

$$678 = 6 \cdot 10^2 + 7 \cdot 10^1 + 8 \cdot 10^0$$

and so on.

In general, any integer can be uniquely expressed in the form $a_n 10^n + a_{n-1} 10^{n-1} + \cdots + a_2 10^2 + a_1 10^1 + a_0$ for some n, where each digit a_i satisfies $0 \le a_i \le 9$ for $i = 0, 1, \ldots, n$. This can be expressed as $a_n a_{n-1} \ldots a_2 a_1 a_0$. In short,

$$a_n a_{n-1} \ldots a_2 a_1 a_0 = \sum_{k=0}^{n} a_k \cdot 10^k \tag{1.5.1}$$

10 is said to be the base of this system and we may write $(a_n a_{n-1} \cdots a_1 a a_0)_{10}$ to emphasize the base.

Another important system is the BINARY system. This system is significant because it is used in computers. Every nonnegative integer has a binary representation similar to Eq. (1.5.1) with powers of two replacing powers of 10.

$$b_m b_{m-1} \ldots b_2 b_1 b_0 = \sum_{k=0}^{n} b_k \cdot 2^k \qquad b_k = 0 \text{ or } 1. \tag{1.5.2}$$

The digits b_k are the binary bits, and are thus called BITS.

We shall write the left side of expression (1.5.2) also as

$$(b_m b_{m-1} \ldots b_2 b_1 b_0)_2$$

if there is any ambiguity. We now define a number system for any base.

Definition. If r is a positive integer the BASE-r REPRESENTATION of the whole number t is given by the $m + 1$ integers $b_0, \ldots, b_m$ such that

$$t = \sum_{k=0}^{m} b_k r^k, \text{ and}$$

$$0 \le b_i \le r \text{ for } 0 \le i \le m, \text{ and}$$

$$b_m \ne 0.$$

The base-r representation of t is written as $t = (b_m b_{m-1} \cdots b_2 b_1 b_0)_r$.

We will look at base-r where $r = 2$ (binary), $r = 12$ (duodecimal), and $r = 16$ (hexadecimal). Hexidecimal problems are in Problem 4.

1.5.1 Conversion Between Binary and Decimal

It is helpful to have a systematic way of converting from binary to decimal and vice-versa. It is easier going from binary to decimal using a method which looks like "synthetic division" or "Horner's Method" from high school algebra.

Example 19. Write $(10111)_2$ in decimal notation.

Solution. Expressing $(10111)_2$ using Eq. (1.5.2),

$$(10111)_2 = 1 \cdot 2^4 + 0 \cdot 2^3 + 1 \cdot 2^1 + 1 \cdot 2^0$$
$$= 16 + 4 + 2 + 1$$
$$= 23$$

The above calculation can be written in the same style as in "synthetic division". Note that in the following table we add each number in a column and then double that total and move it to the next column.

$2|$	1	0	1	1	1
		2	4	10	22
	1	2	5	11	23

//

Exercise 15. Write $(10101)_2$ and $(11110)_2$ in decimal notation.

1.5.2 Conversion from Decimal to Binary

Example 20. Write the base-10 number 45 in binary notation

Solution. Take the highest power of 2 that is less than or equal to the given number, 45. This is 32. Write $45 = 32 + 13$. Take the highest power of 2 that is less than or equal to 13; this is 8. Write $45 = 32 + 8 + 5$. Repeat until we obtain 45 as a sum of powers of 2. Thus $45 = 32 + 8 + 4 + 1 = 2^5 + 2^3 + 2^2 + 2^0$. Hence, $45 = (101101)_2$. //

Alternative Solution There is a different method involving a repeated division algorithm which gives a different way of thinking of the solution to Example 20.

Divide 45 by 2	$45 = 22 \cdot 2 + 1$
Divide the previous quotient, 22, by 2	$22 = 11 \cdot 2 + 0$
Divide the previous quotient, 11, by 2	$11 = 5 \cdot 2 + 1$

Divide the previous quotient, 5, by 2 $\qquad$ $5 = 2 \cdot 2 + 1$

Divide the previous quotient, 2, by 2 $\qquad$ $2 = 1 \cdot 2 + 0$

Stop when the previous quotient is 1

Obtain the digits for the binary number by reading from bottom up, starting with the last quotient, and then the remainder, which gives $(101101)_2$. The above procedure can be rewritten as follows:

$$2\underline{|45}$$
$$2\underline{|22} \ldots 1$$
$$2\underline{|11} \ldots 0$$
$$2\underline{|5} \ldots 1$$
$$2\underline{|2} \ldots 1$$
$$1 \ldots 0$$

$/\!/$

Exercise 16. A. Write the base-ten number 451 in binary notation.
B. Write the base-ten number 5001 in binary notation.

1.5.3   Arithmetic in Binary Systems

The operation in binary systems is essentially the same as in decimal systems. We shall discuss addition, multiplication, subtraction, and division.

Addition of Binary Numbers

The addition in base-2 is constructed using the following facts. $0+0 = 0, 0+1 = 1$, $1 + 0 = 1$, and $1 + 1 = 10$ (that is, write 0 and carry the 1) i.e.

+	0	1
0	0	1
1	1	10

This means we could say "$1 + 1 = 10$ because we carry a 1" in base-2!

Example 21. Add $(10110)_2$ and $(1010)_2$.

Solution.

$$
\begin{array}{cccccc}
 & 1 & 0 & 1 & 1 & 0 \\
+ &   & 1 & 0 & 1 & 0 \\
\hline
1 & 0 & 0 & 0 & 0 & 0
\end{array}
$$

In binary notation, $(10110)_2 + (1010)_2 = (100000)_2$.   //

Example 22. Find $(101101)_2 + (1100)_2$.

Solution.

$$
\begin{array}{cccccc}
1 & 0 & 1 & 1 & 0 & 1 \\
+ &   & 1 & 1 & 0 & 0 \\
\hline
1 & 1 & 1 & 0 & 0 & 1
\end{array}
$$

So $(101101)_2 + (1100)_2 = (111001)_2$.   //

Exercise 17. Evaluate

(a) $(11011)_2 + (11001)_2$
(b) $(110111)_2 + (1100)_2 + (110111)_2$

Multiplication of Binary Numbers

The rule for decimal multiplication also holds for the multiplication of binary
numbers. It is based on the following facts: $0 \cdot 0 = 0, 0 \cdot 1 = 0, 1 \cdot 0 = 0 \ 1 \cdot 1 = 1$

$$
\text{i.e.} \quad
\begin{array}{c|cc}
\times & 0 & 1 \\
\hline
0 & 0 & 0 \\
1 & 0 & 1
\end{array}
$$

Example 23. Calculate $\big((101)_2\big)^2$.

Solution. We need to write out $(101)_2 \cdot (101)_2$. Since $(101)_2 = 2^2 + 1$,

$$
\big((101)_2\big)^2 = (2^2 + 1)(2^2 + 1) = 2^4 + 2 \cdot 2^2 + 1 = 2^4 + 2^3 + 1
$$
$$
= (11001)_2.
$$
                                                                             //

Binary multiplication is done this way (called EXPANDED NOTATION) in elementary school mathematics. We can rewrite the above computation in more compressed
form (called the FINAL ALGORITHM) as follows. Note the "carrying" from the third
column to the fourth.

$$
\begin{array}{ccccc}
 &   &   & 1 & 0 & 1 \\
\times &   &   & 1 & 0 & 1 \\
\hline
 &   &   & 1 & 0 & 1 \\
 &   & 0 & 0 & 0 &   \\
+ & 1 & 0 & 1 &   &   \\
\hline
 & 1 & 1 & 0 & 0 & 1 \\
\end{array}
$$

Exercise 18. Evaluate

(a) $(10110)_2 \cdot (1011)_2$
(b) $(101011)_2 \cdot (1011)_2$

Subtraction in the Binary System

Recall that subtraction is the inverse operation of addition. Therefore we have: $0 - 0 = 0, 1 - 0 = 1$, and $1 - 1 = 0, 0 - 1 = 1$ (with borrowing 1 from the next column if possible) i.e.

$$
\begin{array}{c|cc}
- & 0 & 1 \\
\hline
0 & 0 & 1 \\
1 & 1 & 0 \\
\end{array}
\quad \text{(borrowing 1 from the next column).}
$$

Example 24. Evaluate $(1101)_2 - (1011)_2$.

Solution.
$$
\begin{array}{cccc}
 & \overset{0}{1} & \overset{2}{\cancel{1}} & \overset{}{\cancel{0}} & 1 \\
- & & 1 & 0 & 1 & 1 \\
\hline
 & & 0 & 0 & 1 & 0 \\
\end{array}
$$
So, $(1101)_2 - (1011)_2 = (10)_2$.   //

Example 25. Find $(11101)_2 - (10011)_2$.

Solution.
$$
\begin{array}{ccccc}
 & 1 & 1 & \overset{0}{\cancel{1}} & \overset{2}{\cancel{0}} & 1 \\
- & 1 & 0 & 0 & 1 & 1 \\
\hline
 & 0 & 1 & 0 & 1 & 0 \\
\end{array}
$$
So, $(11001)_2 - (10011)_2 = (1010)_2$.   //

Exercise 19. Evaluate $(111101)_2 - (110111)_2$ and $(101101)_2 - (1111)_2$.

Division in the Binary System

Division in base-2 will use the same long division procedure as we used in base-10.

Example 26. What is $(11101)_2 \div (101)_2$?

Solution.

$$
\begin{array}{r}
101 \\
101\,)\overline{11101} \\
101 \\
\hline
1001 \\
101 \\
\hline
100
\end{array}
$$

Thus $(11101)_2 \div (101)_2 = (101)_2$ as quotient with remainder $(100)_2$.   //

Exercise 20. Evaluate

A. $(111111)_2 \div (101101)_2$.
B. $(111111010010)_2 \div (101101)_2$.

1.5.4   Duodecimal Number System

If $b = 12$ then the number system is called the DUODECIMAL system. Again the digits involved are elements of the set $\{0, 1, 2, 3, 4, 5, 6, 7, 8, 9, T, E\}$. In duodecimal, T denotes 10 and E denotes 11.

Conversion from Decimal to Duodecimal System

Divide n and each succeeding quotient by 12 until a zero quotient is obtained. The sequence of remainders in reverse order yields the duodecimal representation of n.

Example 27. Let $n = 275$ in base 10. What is the number of the duodecimal representative?

$$275 = 22(12) + 11$$
$$22 = 1(12) + 10$$
$$1 = 0(12) + 1$$
$$0 = 0(12) + 0$$

Hence $n = 275 = 1 \cdot 12^2 + 10 \cdot 12 + 11 \cdot 12^0 = 1TE$.

Example 28. If n is 160 in base 10, what is it in base 12?

$$160 = 13(12) + 4$$
$$13 = 1(12) + 1$$
$$1 = 0(12) + 1$$
$$0 = 0(12) + 0$$

Hence $n = 160 = 1 \cdot 12^2 + 1 \cdot 12 + 4 \cdot 12^0 = 114_{(12)}$.

Conversion from Duodecimal to Decimal System

Given an integer n with duodecimal representation

$$n = (a_k a_{k-1} \ldots a_0)_{12},$$

its decimal representation can be obtained by computing

$$n = 12^k a_k + 12^{k-1} a_{k-1} + \ldots + 12^1 a_1 + a_0.$$

Example 29. Let $n = 114_{12}$. Then $n = 1 \cdot 12^2 + 1 \cdot 12^1 + 4 = 144 + 12 + 4 = 160_{10}$.

Example 30. Let $n = (3T7)_{12}$. Then $n \cong 3 \cdot 12^2 + 10 \cdot 12^1 + 7 \cdot 12^0 = 432 + 120 + 7 = 559$.

Exercise 21. Convert the following Decimal numbers to their Duodecimal equivalents and vice versa.

  i. 523
 ii. 6,234
iii. $(2E4)_{12}$
 iv. $(T083)_{12}$

Problem Set 1.5:

1. Convert the following Binary numbers to Duodecimal numbers and vice versa.

   (a) $(1110110)_2$
   (b) $(1010101)_2$
   (c) $(23T)_{12}$
   (d) $(E5)_{12}$

2. Complete the binary sums and difference without first converting to Decimal equivalents.

   (a) $(1001)_2 + (1101)_2$
   (b) $(101)_2 + (111)_2$
   (c) $(10011)_2 - (1101)_2$
   (d) $(101010)_2 - (11001)_2$

3. Complete the binary products and divisions without first converting to Decimal equivalents.

   (a) $(1101)_2 \cdot (1011)_2$
   (b) $(101)_2 \cdot (111)_2$
   (c) $(11011)_2 \div (11)_2$
   (d) $(1010101)_2 \div (101)_2$

4. Another commonly used number system is the Hexadecimal system. This system works in base 16 and uses the symbols

$$\{0, 1, \ldots, 8, 9, A, B, C, D, E, F\}.$$

   (a) Convert the following Decimal numbers to their Hexadecimal equivalents and vice versa.
      (i) 309
      (ii) 3471
      (iii) $(52)_{16}$
      (iv) $(DFE)_{16}$
      (v) $(ACE123)_{16}$
   (b) Complete the Hexadecimal sums and differences without first converting to Decimal equivalents.
      (i) $(28)_{16} + (34)_{16}$
      (ii) $(8C)_{16} + (A7)_{16}$
      (iii) $(34)_{16} - (28)_{16}$
      (iv) $(349)_{16} - (2F1)_{16}$
      (v) $(FED)_{16} - (CBA)_{16}$

Chapter 2
Greatest Common Divisors, Diophantine Equations, and Combinatorics

Introduction This chapter deals with the structure of how prime numbers are put together (Prime Factorization) and the tools of number theory which include greatest common divisors (GCD), least common multiples (LCM). These tools provide the beginning of important theorems such as the Fundamental Theorem of Arithmetic and the Euclidean Algorithm and its byproduct. Problems which have only solutions in integers (Diophantine equations) are discussed. Application to combinatorics is also mentioned.

2.1 GCD and LCM Through the Fundamental Theorem of Arithmetic

Overview In this section, we use the Fundamental Theorem of Arithmetic (prime factorization), Theorem 1, to find the greatest common divisor (GCD) and the least common multiple (LCM) of two integers. We will first use the unique prime factorization to compute the GCD, then define the least common multiple (LCM) of two integers and give two ways of computing the LCM. One will be through prime factorization, one will be through its relationship to the GCD, and the third will be an algorithm similar to that for calculating the GCD. This approach is the one with which you are probably familiar, as it is a part of the mathematics of elementary school. It is covered in the first semester of a content based math course for future elementary and middle school teachers.

We continue in this section with the idea of divisibility, but instead of asking whether a given integer a is divisible by the integer d, we look at whether two integers a and b are both divisible by d. We are especially interested in the largest integer which is a divisor of a and also a divisor of b. That integer will be called the GREATEST COMMON DIVISOR of a and b. It proves very useful in problem solving, showing that there are real numbers that aren't fractions (are irrational.)

© Springer International Publishing Switzerland 2015
R.S. Millman et al., *Problems and Proofs in Numbers and Algebra*,
DOI 10.1007/978-3-319-14427-6_2

The formal definition of this concept and its first applications are the contents of this section. In the next section we will then use the Division Algorithm to find a very efficient method (the Euclidean Algorithm) to find the greatest common divisor of two numbers.

Definition. Let a and b be integers neither of which is zero. If $c \in \mathbb{N}$, c divides a and c divides b then c is called a COMMON DIVISOR of a and b. A positive integer, d, is the GREATEST COMMON DIVISOR of a and b, written $d = \mathrm{GCD}(a, b)$, provided that

1. d is a common divisor of a and b and
2. if c is any common divisor of a and b, then $c \leq d$.

If $n \in \mathbb{Z}$ then the set of positive DIVISORS of n is the set of elements

$$d(n) = \{c \in \mathbb{N} \mid c \text{ divides } n\}.$$

Note that while a divisor of n may be negative, the greatest common positive divisor of two numbers must be positive by definition.

Example 31. What is the greatest common divisor of 24 and 60?

Solution. We collect the positive divisors of 24 and 60 in the sets:

$$d(24) = \{1, 2, 3, 4, 6, 8, 12, 24\}$$

and

$$d(60) = \{1, 2, 3, 4, 5, 6, 10, 12, 15, 20, 30, 60\}.$$

The positive common divisors of 24 and 60 is the set of elements in both $d(24)$ and $d(60)$; that is, $d(24) \cap d(60)$. Since

$$d(24) \cap d(60) = \{1, 2, 3, 4, 6, 12\}$$

and the largest common element of these is 12, the GCD(24, 60) is 12.   //

Exercise 22. Find the greatest common divisor of 36 and 118.

The way in which we found GCD(24, 30) is called the **method of exhaustion**. While it is not efficient in finding the GCD of two numbers (for example, try to find the GCD(136, 62) with this method and compare with your efforts in Example 38 when we get there), it does show that there is a GCD(a, b) for any numbers a and b. Here is the reasoning: Continuing to use the notation $d(n)$ for the set of divisors of n, we note that both $d(a)$ and $d(b)$ are finite sets and they have the integer 1 in common. Therefore their intersection has a largest element. This element is the greatest common divisor of a and b and is at least 1. Symbolically,

$$1 \leq \mathrm{GCD}(a, b) = \text{largest element of } (d(a) \cap d(b)) \leq \text{minimum}\{|a|, |b|\}.$$

Let's look again at Example 31 but see what happens if we use the Fundamental Theorem of Arithmetic to factor both 24 and 60 as the product of primes.

Example 32. Using prime factorization, find the $GCD(24, 60)$.

Solution. We note that

$$24 = 2^3 \cdot 3 \text{ and } 60 = 2^2 \cdot 3 \cdot 5 \qquad (2.1.1)$$

where we have purposely written the primes in increasing order. Looking at the prime factorization, Eq. (2.1.1), we see that the largest integer which divides both is

$$d = 2^2 \cdot 3 = 12 = GCD(24, 60).$$

$\textit{//}$

Thus we have found the $GCD(24, 60)$ by factoring both numbers into primes and then seeing which primes are in <u>both</u> the factorizations. In this case, they are 2 and 3 but not 5. We then see what is the largest exponent of these primes that appears in each of them (exponent of 2 for the prime 2 and 1 for the prime 3). The same procedure always works, although the notation gets to be complicated as we see in Theorem 3.

Theorem 3 (GCD via Prime Factorization). *Let a and b be positive integers and suppose that the prime factors of a are $p_1, \ldots p_k$ and the prime factors of b are $q_1, \ldots q_l$. We then write*

$$a = p_1^{s_1} \ldots p_k^{s_k} \text{ and } b = q_1^{r_1} \ldots q_l^{r_l}$$

where each of the exponents is positive. If the primes are re-arranged so that the first v of them are equal ($p_1 = q_1, \ldots p_v = q_v$), then the $GCD(a,b) = p_1^{t_1} \ldots p_v^{t_v}$ where t_i is the minimum of r_i and s_i.

To show the theorem, we need the following lemma.

Lemma 1. *Let $n = p_1^{n_1} p_2^{n_2} \cdots p_t^{n_t}$ with $n_i > 0$ for each i and let m be a positive integer. Then $m \mid n$ if and only if $m = p_1^{m_1} p_2^{m_2} \cdots p_t^{m_t}$ with $m_i \leq n_i$ for each i.*

Proof. If $m = p_1^{m_1} p_2^{m_2} \cdots p_t^{m_t}$ with $m_i \leq n_i$ for each i, then

$$n = p_1^{n_1} \cdots p_t^{n_t} = (p_1^{n_1 - m_1} \cdots p_t^{n_t - m_t})(p_1^{m_1} \cdots p_t^{m_t}). \qquad (2.1.2)$$

Since $n_i \geq m_i$, we have $p_1^{n_1 - m_1} \cdots p_t^{n_t - m_t} \geq 1$ and $m \mid n$.

Conversely if $m \mid n$, then $n = mk$ for some integer k. Since $n = p_1^{n_1} \cdots p_t^{n_t}$ and the prime power decomposition of n is unique (Corollary 1), we have the prime power decomposition of m and k must be the same as those in n. Therefore, we have

$$m = p_1^{m_1} \cdots p_t^{m_t} \text{ and } k = p_1^{k_1} \cdots p_t^{k_t}, \quad m_i \geq 0, k_i \geq 0. \qquad (2.1.3)$$

Since $n = mk$, we have $n_i = m_i + k_i$ and then $n_i \geq m_i$. ∎

Now we are ready to prove Theorem 3.

Proof. It suffices to show that if $a = p_1^{s_1} p_2^{s_2} \cdots p_v^{s_v}$ and $b = p_1^{r_1} p_2^{r_2} \cdots p_v^{r_v}$, then $\text{GCD}(a, b) = p_1^{t_1} p_2^{t_2} \cdots p_v^{t_v}$ where t_i is the minimum of r_i and s_i. Let $d = p_1^{t_1} p_2^{t_2} \cdots p_v^{t_v}$. Since t_i is the minimum of r_i and s_i, $t_i \leq r_i$ and $t_i \leq s_i$. Therefore, by the above lemma, $d \mid a$ and $d \mid b$. Suppose now $g \mid a$ and $g \mid b$. Then by the lemma above again, $g = p_1^{g_1} p_2^{g_2} \cdots p_v^{g_v}$ with $g_i \leq s_i$ and $g_i \leq r_i$ for each i. But since t_i is the minimum of s_i and r_i, this implies that $g_i \leq t_i$. Therefore, again by the above lemma, $g \mid d$. Thus, $d = \text{GCD}(a, b)$. ∎

Remark. The above lemma provides an easy way to find all the positive divisors of a positive integer once its prime power decomposition is obtained.

Example 33. Find all positive divisors of 108.

Solution. Since $108 = 2^2 \times 3^3$, the divisors of 108 are

$$1 \cdot 1 \quad 1 \cdot 3 \quad 1 \cdot 3^2 \quad 1 \cdot 3^3$$
$$2 \cdot 1 \quad 2 \cdot 3 \quad 2 \cdot 3^2 \quad 2 \cdot 3^3$$
$$2^2 \cdot 1 \quad 2^2 \cdot 3 \quad 2^2 \cdot 3^2 \quad 2^2 \cdot 3^3$$

or

$$1 \cdot 1 \quad 1 \cdot 2 \quad 1 \cdot 2^2$$
$$3 \cdot 1 \quad 3 \cdot 2 \quad 3 \cdot 2^2$$
$$3^2 \cdot 1 \quad 3^2 \cdot 2 \quad 3^2 \cdot 2^2$$
$$3^3 \cdot 1 \quad 3^3 \cdot 2 \quad 3^3 \cdot 2^2$$

     //

Example 34. Find the $GCD(504, 3675)$ using the prime factorization method.

Solution. It is easy to find that

$$504 = 2^3 \cdot 3^2 \cdot 7 \text{ and } 3675 = 3^1 \cdot 5^2 \cdot 7^2.$$

Thus, since the only primes that are in both numbers are 3 and 7 and each has exponent 1 in one of the numbers, the theorem says that

$$GCD(504, 3675) = 3 \cdot 7 = 21.$$

     //

Recall that an integer greater than 1 is prime if its only positive divisors are itself and one. Of course, a prime number p is also divisible by $-p$ and -1. By analogy, a very appealing situation will be when, given two integers, the only common positive divisor they have is 1. We formalize that concept in the next definition which will be quite important for the remainder of this book.

Definition. The integers a and b are RELATIVELY PRIME if the $\text{GCD}(a, b) = 1$.

Being relatively prime is one extreme relationship for a pair of integers. The next exercise is concerned with another extreme relationship.

Exercise 23. If $a, b \in \mathbb{Z}$ and $\text{GCD}(a, b) = b$, what is the relationship between a and b? (Prove that your answer is correct.)

We now define rational and irrational numbers and apply the ideas above to show that the $\sqrt{2}$, $\sqrt{3}$ and $\sqrt{5}$ are irrational (in fact, the same logic can show that $\sqrt{p}$ is irrational if p is a prime number). This is another example of problem solving using an established foundation (in this case, the concept of a GCD.)

Definition. If x is a real number, $x = \frac{r}{s}$ where $r, s \in \mathbb{Z}$ and $s \neq 0$, then x is called a RATIONAL NUMBER. We say x is an IRRATIONAL NUMBER if it is not a rational number.

Proposition 3 (Reduction to Lowest Terms). *If x is a non-zero rational number, then x can be written as*

$$x = \frac{a}{b}, \text{ where } a \text{ and } b \text{ are relatively prime integers.} \tag{2.1.4}$$

Proof. Suppose x is rational so that $x = \frac{r}{s}$ for $r, s \in \mathbb{Z}$. Let $d = \text{GCD}(r, s)$. Then

$$x = \frac{r}{s} = \frac{\frac{r}{d}}{\frac{s}{d}}.$$

If we let $a = \frac{r}{d}$ and $b = \frac{s}{d}$ then both are integers, since d is a divisor of both and are relatively prime (why?). Thus, we have written the integer x as in Eq. (2.1.4). ∎

The ability to represent any rational integer as the quotient of two relatively prime integers ("reduce to lowest terms"), as in (2.1.4), allows us to prove results by the technique called "proof by contradiction". The following theorem uses this technique as do Theorem 9 and the Rational Root Theorem (Theorem 20).

Theorem 4. *The number $\sqrt{2}$ is an irrational number.*

Proof. To prove that $\sqrt{2}$ is irrational, we will assume the opposite (i.e. that $\sqrt{2}$ is rational) and show that assumption leads to a contradiction and so is incorrect. Since the assumption that $\sqrt{2}$ is rational is false, it must be that $\sqrt{2}$ is irrational, which is what we are trying to prove. This technique is called "proof by contradiction".

Assume for the moment that the $\sqrt{2}$ is rational and so it is the quotient of two integers. Thanks to (2.1.4), we may assume that

$$\sqrt{2} = \frac{a}{b}, \text{ where } \text{GCD}(a, b) = 1. \tag{2.1.5}$$

This equation means that $a = b\sqrt{2}$ or

$$a^2 = 2b^2 \tag{2.1.6}$$

and so a^2 is even. If a were odd (say $a = 2\ell + 1$), its square ($4\ell^2 + 4\ell + 1 = 2(2\ell^2 + 2\ell) + 1$) would also be odd. Therefore, a is even which means that $a = 2k$ for some $k \in \mathbb{N}$. Putting $a = 2k$ into (2.1.6) yields

$$4k^2 = 2b^2.$$

Thus b^2, and hence b, is even by the same argument. Since both a and b are even, each is divisible by 2. Thus, $\text{GCD}(a,b) \geq 2$, which contradicts our assumption [Eq. (2.1.5)] and the proof is finished. ∎

We now turn our attention briefly to the concept of least common multiple (LCM) of two integers. While the GCD will be much used in the remainder of this chapter, we will treat the LCM briefly as it will not play as much of a part in the remainder of this text. However, both the GCD and the LCM are very much a part of the mathematics of elementary schools.

Definition. The LEAST COMMON MULTIPLE of two integers a and b, $LCM(a,b)$, is the smallest positive integer that is a multiple of both a and b.

We can recast this definition to be similar to the definition of $GCD(a,b)$ by saying that $l = LCM(a,b)$ if

1. l is a multiple of a and of b
2. if k is any other multiple of a and b then $l \leq k$.

If $n \in \mathbb{Z}$ then the POSITIVE MULTIPLES of n are collected in the set:

$$m(n) = \{c \in \mathbb{N} | n \text{ divides } c\}.$$

The $LCM(a,b)$ is the smallest integer in $m(a) \cap m(b)$.

Note that all elements of $m(a)$ and $m(b)$ are positive and if a and b are either both positive or both negative, then $m(a)$ and $m(b)$ share at least the element ab and thus there is a $LCM(a,b)$—that is, $m(a) \cap m(b) \neq \emptyset$.

Example 35. Find the least common multiple of 270 and 630.

Solution. We will use the prime factorization of 270 and 630 to find the $LCM(270, 630)$. It is easy to show that

$$270 = 2 \cdot 3^3 \cdot 5 \text{ and } 630 = 2 \cdot 3^2 \cdot 5 \cdot 7.$$

Any multiple of both 270 and 630 must have in its prime factorization at least one 2, three 3s, one 5, and one 7 in it. Thus, the smallest multiple of both numbers must be

$$LCM(270, 630) = 2 \cdot 3^3 \cdot 5 \cdot 7 = 1890.$$

//

It isn't difficult, from looking at this example to prove the following theorem which is analogous to Theorem 3 which uses prime factorization to find the GCD of two integers. Once again, the difficult part of the proof is keeping track of the notation.

Theorem 5 (LCM via Unique Prime Factorization). *Let a and b be positive integers and suppose that the prime factors of a are $p_1 \ldots p_k$ and the prime factors of b are $q_1 \ldots q_l$. We then write*

$$a = p_1^{s_1} \ldots p_k^{s_k} \text{ and } b = q_1^{r_1} \ldots q_l^{r_l}$$

where each of the exponents is positive. If the first v of the primes are equal ($p_1 = q_1, \ldots p_v = q_v$) and t_j is the maximum of r_j and s_j for $1 \leq j \leq v$ then

$$LCM(a,b) = p_1^{t_1} \ldots p_v^{t_v} p_{v+1}^{s_{v+1}} \ldots p_k^{s_k} q_{v+1}^{r_{v+1}} \ldots q_l^{r_l}.$$

We will not include a proof of the fact but ask the reader to supply one for a number of cases in which the notation is easier.

Proposition 4. *If a and b are non-zero integers and $d = GCD(a,b)$ then there are relatively prime integers h and k with*

$$a = dh \text{ and } b = dk \tag{2.1.7}$$

Proof. Since d is the GCD of a and b, $d \,|\, a$ and $d \,|\, b$, so that there are integers h and k which satisfy Eq. (2.1.7). We will now show that, in fact, h and k are relatively prime.

Define $r = GCD(h,k)$. If $r = 1$, then, by definition, h and k are relatively prime and we are done. Note that $\frac{h}{r}$ and $\frac{k}{r}$ are integers and $a = (rd) \left(\frac{h}{r}\right)$ and $b = (rd) \left(\frac{k}{r}\right)$.

If $r > 1$, then $rd > d$, $rd \,|\, a$, and $rd \,|\, b$ which is a contradiction since $d = GCD(a,b)$. Therefore $r = 1$, so h and k are relatively prime. ∎

Theorem 6. *For any non-zero integers a and b,*

$$GCD(a,b)LCM(a,b) = ab.$$

Proof. Let $GCD(a,b) = d$. Then there exist nonzero positive integers u and v where $a = du$ and $b = dv$ and $GCD(u,v) = 1$. Then we see that $LCM(a,b) = uvd$ and:

$$GCD(a,b) \cdot LCM(a,b) = d \cdot (uvd) = (du)(dv) = ab.$$

∎

Theorem 6 provides a direct link between the GCD and LCM of two arbitrary positive integers. Thus given either the GCD or LCM, one is able to find the other by a simple division of integers. However there is a method which simultaneously determines the GCD and LCM of two positive integers using unique factorization; we call the following method "simultaneously determining GCD and LCM".

Example 36. Determine $LCM(40, 155)$ given that the $GCD(40, 155) = 5$.

Solution. By Theorem 6, we see that:

$$LCM(40, 155) = \frac{40 \times 155}{GCD(40, 155)} = \frac{6200}{5} = 1240 = 5 \times 8 \times 31.$$

If we do not assume that the $GCD(40, 155) = 5$, we can still determine the LCM quickly. We notice from the divisibility tests that 5 is a common divisor of 40 and 155. We create the following table displays the quotients of dividing both integers by 5.

$$
\begin{array}{c|cc}
5 & 40 & 155 \\
\hline
 & 8 & 31
\end{array}
$$

Since 8 and 31 are relatively prime, we see that $5 = GCD(40, 155)$. Also we can read the *LCM* off of the table by multiplying the divisor by the relatively prime quotients. Hence $LCM(40, 155) = 5 \times 8 \times 31$.   //

This process can be repeated in the case we do not guess the *GCD* for our first divisor. Recall Example 34 where we found $GCD(504, 3675) = 21$. We now produce an alternative solution to this example using the "simultaneously determining GCD and LCM" method.

Solution (Alternative Solution to Example 34). Using the divisibility tests we first recognize that 3 is a common divisor to 504 and 3675. After division by 3 we recognize that 7 is the smallest prime that divides the quotients.

$$
\begin{array}{c|cc}
3 & 504 & 3675 \\
7 & 168 & 1225 \\
\hline
 & 24 & 175
\end{array}
$$

Since 24 and 175 are relatively prime, we have determined $GCD(504, 3675) = 21$. We have additionally determined:

$$LCM(504, 3675) = 3 \times 7 \times 24 \times 175 = 2^3 \times 3^2 \times 5^2 \times 7^2 = 88200.$$ //

Exercise 24. Using the "simultaneously determining GCD and LCM" method, determine the LCM of 270 and 630.

The remainder of this section is needed only in an exercise in Sect. 2.3 but not in the material of this text. It indicates how to extend these ideas to more than two integers.

Definition. If a, b, c are integers (not all zero) then the GREATEST COMMON DIVISOR of a, b, and c is the largest integer which divides all three. We will write $GCD(a, b, c)$ for the greatest common divisor.

Example 37. What is the GCD(12, 18, 30)?

Solution. Since the numbers 12, 18, and 30 are relatively small we can easily write down all of their divisors individually. A good way to organize this information is in the form of a chart.

Number	Divisors
12	1, 2, 3, 4, 6, 12
18	1, 2, 3, 6, 9, 18
30	1, 2, 3, 5, 6, 10, 15, 30

From the chart, we see that GCD(12, 18, 30) = 6. //

Exercise 25. Prove that for any integers a, b, c not all zero, $GCD(a, b, c)$ exists.

Remark. Unfortunately the *GCD* and *LCM* of three positive integers a, b, c do not behave as nicely as the case we considered only two integers. To be specific, in general:

$$GCD(a, b, c)LCM(a, b, c) \neq abc.$$

Exercise 26. Find an example of three distinct integers a, b, c such that

$$GCD(a, b, c) \cdot LCM(a, b, c) \neq abc.$$

Problem Set 2.1:

1. Find the $GCD(14, 22)$ by both the method of exhaustion and via prime factorization.
2. Find the $GCD(600, 28000)$ by both the method of exhaustion and via prime factorization. Which is easier?
3. Find the $LCM(14, 22)$ via prime factorization.
4. Find the $LCM(600, 28000)$ via prime factorization.
5. Show that $GCD(a, b)LCM(a, b) = ab$ in the cases below (see Problems 1–4 above).

   (a) $a = 14$ and $b = 22$
   (b) $a = 600$ and $b = 28000$

6. Prove that $\sqrt{3}$ is irrational. (Hint: Look at the proof of Theorem 4.)
7. Prove that $\sqrt{5}$ is irrational.
8. If r is a rational number, prove that $r\sqrt{5}$ is irrational.
9. If r is an irrational number, is it necessarily true that $r\sqrt{5}$ is irrational? If this is a true statement, then prove it. If it is false provide a counter example.

10. Let n be between 1000 and 1200 inclusively. Assume that if n is divided by 7, 9, 11, the remainders are 4, 6, 8 respectively. What is the smallest value of n?

11. Let n be between 9000 and 10,000 inclusively. Assume that if n is divided by 5, 9, 13, or 17, the remainders are all 3. What is the smallest value of n?

12. John goes to the library every 2 days and Linda goes to the library every 3 days. They met at the library last Sunday. When will they meet again on a Sunday?

13. Every 5 days Hayden eats steak at the town diner. Lucy eats at the same diner every 2 days and if she goes on a Sunday, she brings her friend Janice. If Hayden, Lucy, and Janice all ate at the dinner on Sunday, January 1, when will be the next date all three of them eat there again?

14. Suppose for positive integers a and b, we know $ab = 1200$ and GCD $(a, b) = 10.$

   (a) If $a > b$, what are the possible combinations for a and b?
   (b) For each combination found in Part a, determine $LCM(a, b)$.

15. Suppose that there is only one prime in the factorization of a and only one in the factorization of b so that

$$a = p^s \text{ and } b = q^r \tag{2.1.8}$$

where p and q may be different.

   (a) According to Theorems 3 and 5, what is the $GCD(a, b)$ and the $LCM(a, b)$? (You will need to separate out two cases: $p = q$ and $p \neq q$.)
   (b) Prove Theorems 3 and 5 for the special case of Eq. (2.1.8).

16. Suppose that there are two primes in the factorization of a and two primes in the factorization of b and all four primes are different.

   (a) According to Theorems 3 and 5, what is the $GCD(a, b)$ and $LCM(a, b)$?
   (b) Prove Theorems 3 and 5 for this special case.

17. Determine the GCD and LCM of 150 and 365 first using prime factorization and then using the "simultaneously determining GCD and LCM" method.

18. Determine the GCD and LCM of 1050 and 3276 using the "simultaneously determining GCD and LCM" method.

19. There are 48 oranges, 36 mangos and 60 pears to be distributed in boxes. The distribution mandates that each box contains an equal amount of the same fruit. What is the maximum number of boxes needed and how many oranges, mangos and pears are in each box?

20. Let $a, b, q_1, q_2, q_3 \in \mathbb{N}$ and let

$$a = bq_1 + 4098$$
$$b = 4098q_2 + 582$$
$$4098 = 582q_3 + 24.$$

What is $GCD(a, b)$? (Answer: 6)

2.2   GCD, the Euclidean Algorithm and Its Byproducts

Overview We now turn to the use of the Division Algorithm (see Sect. 1.1), which allows us a more efficient way of finding the GCD. The method about to be presented was discovered by the Greek mathematician Euclid about 2,000 years ago and is called the **Euclidean Algorithm**. We first prove the theorem and then show its usefulness through some examples.

Note that a and $-a$ have the same set of divisors for any integer a. Thus, in computing $GCD(a, b)$ we can assume that a and b are nonnegative as $GCD(a, b) = GCD(|a|, |b|)$.

Theorem 7 (Euclidean Algorithm). *Given $a, b \in \mathbb{N}$ where $b > a$ and $a \nmid b$, apply the division algorithm successively as follows:*

$$b = q_1 a + r_1, 0 < r_1 < a$$

$$a = q_2 r_1 + r_2, 0 < r_2 < r_1$$

$$r_1 = q_3 r_2 + r_3, 0 < r_3 < r_2$$

$$\ldots$$

$$r_{n-2} = q_n r_{n-1} + r_n, 0 < r_n < r_{n-1}$$

$$r_{n-1} = q_{n+1} r_n + 0$$

Then $r_n = GCD(a, b)$. That is, the last non-zero remainder that appears in this manner is equal to $GCD(a, b)$.

The proof of this theorem is based on the following lemma:

Lemma 2. *Let $a, b \in \mathbb{Z}$, and assume that $a > 0$ and a does not divide b. Let q and r be the quotient and remainder respectively of b when divided by a as guaranteed by the Division Algorithm. Then*

$$b = qa + r, \, 0 < r < a \tag{2.2.1}$$

and $GCD(a, b) = GCD(a, r)$.

Proof. Equation (2.2.1) is just the statement of the Division Algorithm so we only need to show that $GCD(a, b) = GCD(a, r)$. Let $d = GCD(a, b)$. To prove that d is also equal to the $GCD(a, r)$, in accordance with the definition of the GCD, we must prove two things:

 (i) d is a divisor of both r and a
 (ii) d is the largest common divisor of r and a. $\tag{2.2.2}$

We first prove Part (i). Now, $d = \text{GCD}(a, b)$ means that d divides b and d divides a. Thus Eq. (2.2.1) shows that d also divides r since

$$d \mid (b - qa) = r.$$

Since d is already a divisor of a, d is a common divisor of r and a, which is what Part (i) requires.

To prove Part (ii), we let $\bar{d}$ be any common divisor of r and a and show that $\bar{d}$ is no larger than d. Certainly, since $\bar{d}$ is a divisor of r and a,

$$\bar{d} \mid r \text{ and } \bar{d} \mid a. \tag{2.2.3}$$

We now show that $\bar{d} \leq d = \text{GCD}(a, b)$ to establish statement (2.2.2). However, Eqs. (2.2.1) and (2.2.3) show that $\bar{d}$ must divide b since

$$\bar{d} \mid (qa + r) = b$$

and $\bar{d} \mid a$. Therefore

$$\bar{d} \leq \text{GCD}(a, b) = d,$$

which means that $\bar{d}$ is no larger than d. Hence Part (ii) [statement (2.2.2)] is established.   ■

We are now ready to prove the theorem.

Proof. Using this lemma we have $\text{GCD}(a, b) = \text{GCD}(a, r_1) = \ldots = \text{GCD}(r_{n-1}, r_n)$ $= \text{GCD}(r_n, 0) = r_n$.   ■

Example 38. Find the greatest common divisor of 136 and 62 using the Euclidean algorithm.

Solution. Since $136 = 2(62) + 12$, the remainder of 136 on division by 62 is 12. We can apply the Euclidean Algorithm and obtain

$$\text{GCD}(136, 62) = \text{GCD}(62, 12).$$

But $62 = 5(12) + 2$ so another application gives (since 2 is the remainder of 62 on division by 12),

$$\text{GCD}(136, 62) = \text{GCD}(62, 12) = \text{GCD}(12, 2) = 2.$$
                                                                                                //

The process of Example 38 can be displayed (with the circled numbers being the multipliers and subtraction made in the vertical columns) by the chart below.

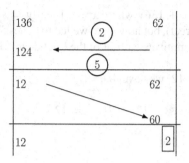

The last line gives the GCD$(12, 2) = 2$ by inspection since 2 divides 12. Note that the GCD is in a box, not a circle.

Exercise 27. Find the GCD$(118, 36)$ using the Euclidean Algorithm as in Example 38.

From the Euclidean Algorithm (Theorem 7) we next gain a wonderful byproduct (Theorem 8). We will first look at some examples which depend on the important observation that the remainder of a on division by b can be written as an integer, x, times a plus another integer, y, times b. This phenomenon is called expressing r as an integer linear combination of a and b. In Example 38, $r = 2$, $a = 136$, $b = 62$, and $2 = 62x + 136y$ where $x = 11$ and $y = -5$. How to obtain $x = 11$ and $y = -5$ is difficult. Right now we are only asking you to do arithmetic so that $2 = 62(11) + 136(-5)$.

We now carefully define the necessary terminology to develop another method for solving GCD problems, mainly backwards substitution. We will repeat Example 38 to show how to obtain $x = 11$ and $y = -5$ and along with Example 40 supports this method before stating and proving in a special case the Euclidean Algorithm Byproduct.

Definition. Let $a, b \in \mathbb{Z}$. r is an INTEGRAL LINEAR COMBINATION of a and b if there are integers x and y such that

$$r = ax + by. \tag{2.2.4}$$

We will usually just write that r is a linear combination of a and b unless there is some ambiguity.

Example 39. Recall from Example 38 that GCD$(136, 62) = 2$. Using some arithmetic insight (guessing) we see

$$2 = 62(11) + 136(-5)$$

which shows Eq. (2.2.4) to be true in this case with $x = 11$ and $y = -5$ and $r = $ GCD$(136, 62) = 2$. Thus, we are able to show that $2 = $ GCD$(136, 62)$ is a linear combination of 136 and 62 because we were lucky and realized that

$$2 = 62x + 136y \text{ if } x = 11 \text{ and } y = -5. \tag{2.2.5}$$

Showing that $2 = 62x + 136y$ where $x = 11$ and $y = -5$ is all well and good (and mathematically correct), but how were we led to "guess" $x = 11$ and $y = -5$? Our method (which will motivate a proof of the Euclidean Algorithm Byproduct) is as follows.

Since $136 = 2(62) + 12$, we have that

$$136 - 2(62) = 12. \tag{2.2.6}$$

But $62 = 5(12) + 2$ means that

$$62 - 5(12) = 2. \tag{2.2.7}$$

We now go "backwards." Putting (2.2.6) into (2.2.7) gives, with $a = 62$ and $b = 136$,

$$2 = 62 - 5(136 - 2(62)) = a - 5(b - 2a)$$
$$= 11a - 5b.$$

Thus $2 = (62)11 + 136(-5)$, as promised, and the $\text{GCD}(a, b)$ is indeed a linear combination of a and b. Using the Euclidean Algorithm Byproduct will mean that we can obtain the linear combination of (2.2.6) with $x = 11$ and $y = -5$ without guessing.

Here's another example which gives insight into how the proof of Theorem 8 should work.

Example 40 (Backwards Substitution). Let $b = 123$ and $a = 36$. Find the $\text{GCD}(123, 36)$ and show that it is a linear combination of 123 and 36.

Solution. The Division Algorithm gives:

$$123 = 3(36) + 15 \tag{A}$$
$$36 = 2(15) + 6 \tag{B}$$
$$15 = 2(6) + 3 \tag{C}$$
$$6 = 2(3) + \boxed{0} \tag{D}$$

so that using the Euclidean Algorithm, Theorem 7,

$$\text{GCD}(123, 36) = \text{GCD}(36, 15) = \text{GCD}(15, 6) = \text{GCD}(6, 3) = 3.$$

The answer for the first part of this example is that $\text{GCD}(123, 36) = 3$.

However, now look at what the chain of substitutions from Eqs. (A), (B) and (C) gives! We will now show that these three equations give the "integral linear combination" result which appears is the Euclidean Algorithm Byproduct, Theorem 8. A simple "moving to the other side" process on (A), (B) and (C) yields:

$$123 - 3(36) = 15 \text{ or } b - 3a = 15 \tag{A'}$$

$$36 - 2(15) = 6 \tag{B'}$$

$$15 - 2(6) = 3. \tag{C'}$$

Starting at (C') and using (B') with $b = 123$, $a = 36$ gives

$$GCD(123, 36) = 3 = 15 - 2(36 - 2(15))$$

$$= 15 - 2a + 4(15) = 5(15) - 2a.$$

Now, continuing going backwards up to (A') to replace 15 by $123 - 3(36) = b - 3a$, gives:

$$3 = 5(123 - 3(36)) - 2a = 5b - 15a - 2a.$$

Therefore:

$$GCD(a, b) = GCD(123, 36) = 3 = 5b + (-17)a$$

and so the $GCD(a, b)$ is a linear combination of a and b.   //

   In number theory, the process of Example 40 is (sometimes) called BACKWARDS SUBSTITUTION. Look at the pattern in Eqs. (A), (B), (C), and (D)—the three 15's run diagonally, then the three 6's, then the two 3's and note that 15, 6, and 3 are remainders.

   Backwards substitution can be displayed in a tableau (TABLEAU METHOD) which is more effective than the classical way. The construction of the table below ends with the $GCD(a, b)$ expressed as an integral combination of a and b as we see in Example 41.

Example 41 (Tableau Method). Write the $GCD(123, 36)$ as a linear combination of 123 and 36 using a tableau method.

Solution. We will keep track of the operations by writing a table ("tableau") which follows closely Eqs. (A), (B), (C), and (D) above and looks like the table of Example 40. We are writing $b = 123$ and $a = 36$. Note that the integer below the line is the difference of the two above and is Eq. (A). Similarly, the next vertical triple is (B) and the third is (C).

b	123 ③ 36	a
3a	108 ②	
b-3a	15 36	a
	30	2(b-3a)
b-3a	15 ② 6	a-(2(b-3a))=7a-2b
2(7a-2b)	12	
b-3a-2(7a-2b)	|3| 6	=5b-17a

Thus the GCD$(a, b) = 3 = 5b - 17a$. As a check, $3 = 5(123) - 17(36)$.   //

Example 42. Use the classical and Tableau methods for finding the GCD$(13, 5)$ as a linear combination of 13 and 5.

Solution. Classically, using the Euclidean Algorithm three times, one can write

$$13 = 2 \cdot 5 + 3$$
$$5 = 1 \cdot 3 + 2$$
$$3 = 1 \cdot 2 + 1$$
$$2 = 2 \cdot 1$$

and then back substitute. Then we obtain:

$$1 = 3 - 1 \cdot 2$$
$$2 = 5 - 1 \cdot 3$$
$$3 = 13 - 2 \cdot 5$$

Therefore $1 = 3 - 1(5 - 1 \cdot 3) = 3 - 5 + 1 \cdot 3 = 2 \cdot 3 - 5 = 2(13 - 2 \cdot 5) - 5 = 2 \cdot 13 - 5 \cdot 5$.
Note that the $GCD(13, 5) = 1$ from the last line.

The tableau method, with $b = 13$ and $a = 5$, can be constructed from the above equations as:

b	13 ② 5	a
2a	10	
b-2a	3 ① 5	a
	3	b-2a
b-2a	3 ① 2	a-(b-2a)=3a-b
3a-b	2	
2b-5a	|1|	

Thus $1 = 2b - 5a$ which solves GCD$(13, 5) = 1 = 13x + 5y$ with $x = 2$ and $y = -5$.   //

Theorem 8 (Euclidean Algorithm Byproduct). *If a and b are non-zero integers then the GCD(a, b) is an integral linear combination of a and b.*

We will now give a proof of Theorem 8 (Euclidean Algorithm Byproduct) in the case when we need four iterations of the Euclidean Algorithm, which was the situation sketched out in Example 40. A general proof follows in exactly the same manner but the notation gets a bit awkward during the backwards substitution.

Proof (Sketch). Given $a, b \in \mathbb{Z}$, assume that the $GCD(a, b)$ is known and for simplicity assume $b > a > 0$. We use the Euclidean Algorithm to obtain quotients q_1, q_2, q_3 and q_4 and remainders r_1, r_2, r_3 (respectively) in its fourth step (again, this assumption is to simplify the notation, although the same ideas work in the general case). Thus:

$$b = q_1 a + r_1 \qquad\qquad 0 < r_1 < a \qquad\qquad \text{(E)}$$

$$a = q_2 r_1 + r_2 \qquad\qquad 0 < r_2 < r_1 \qquad\qquad \text{(F)}$$

$$r_1 = q_3 r_2 + r_3 \qquad\qquad 0 < r_3 < r_2 \qquad\qquad \text{(G)}$$

$$r_2 = q_4 r_3 + 0. \qquad\qquad\qquad\qquad\qquad\qquad \text{(H)}$$

Note again the diagonal and vertical pattern with the remainders r_1, r_2, and r_3. We know that

$$GCD(a, b) = GCD(a, r_1) = GCD(r_1, r_2) = GCD(r_2, r_3) = r_3.$$

We now rewrite Eqs. (E), (F), and (G) (mimicking Example 40)

$$b - q_1 a = r_1 \qquad\qquad\qquad\qquad \text{(E')}$$

$$a - q_2 r_1 = r_2 \qquad\qquad\qquad\qquad \text{(F')}$$

$$r_1 - q_3 r_2 = r_3. \qquad\qquad\qquad\qquad \text{(G')}$$

Then starting with (G') we perform the backwards substitution, first with (F'):

$$GCD(a, b) = r_3 = r_1 - q_3 r_2 = r_1 - q_3(a - q_2 r_1)$$

$$= (1 + q_3 q_2) r_1 - q_3 a.$$

Now using (E'), we complete the backwards substitution:

$$GCD(a, b) = (1 + q_3 q_2)(b - q_1 a) - q_3 a$$

$$= (1 + q_3 q_2) b + (-q_1 - q_3 - q_1 q_2 q_3) a.$$

Therefore

$$GCD(a, b) = xa + yb$$

where x and y are the integers given by

$$y = (1 + q_2 q_3) \text{ and } x = -(q_1 + q_3 + q_1 q_2 q_3).$$

Thus, the GCD(a, b) can be written as $xa + yb$ where $x, y \in \mathbb{Z}$ are as in the last equation. This linear combination is the conclusion of the Euclidean Byproduct Theorem in the special case when the process stops in its fourth step. ∎

Exercise 28. Write the GCD$(118, 36)$ as an integral linear combination of 36 and 118 (see Exercise 27) by mimicking the proof of Theorem 8.

Given $a, b \in \mathbb{Z}$, the use of Theorem 8 as a method to find integers x and y such that the GCD$(a, b) = ax + by$ can be cumbersome, as we saw in Example 40. Furthermore, there are many possible integer coefficients x and y that work. For example, if x_0 and y_0 is a solution of

$$\text{GCD}(a, b) = x_0 a + y_0 b$$

then, for any $t \in \mathbb{Z}$, $x = x_0 + bt$ and $y = y_0 - at$ are also solutions of Eq. (2.2.4) (See Problem 5 of Problem Set 2.2). We will see more of the characteristics of other solutions in the next section.

Definition. Let $a, b \in \mathbb{Z}$ be given. If x_0 and y_0 are integers such that

$$\text{GCD}(a, b) = x_0 a + y_0 b$$

then (x_0, y_0) is called a PARTICULAR SOLUTION of the integral linear combination. A GENERAL SOLUTION of the integral linear combination of a and b is

$$\{(x, y) \mid x = x_0 + bt \text{ and } y = y_0 - at \text{ for some } t \in \mathbb{Z}\}.$$

The notion of particular and general solutions will play a major role in the next section, Diophantine Equations, and will allow us to solve many problems which require integer solutions.

Here is another technique for finding a particular solution of the integral linear combination.

Example 43. Find a particular solution and general solution of the integral linear combination for the GCD$(611, 235)$.

Solution. Some computation shows that GCD$(611, 235) = 47$. We are looking for integers x and y such that

$$611x + 235y = 47,$$

which can be reduced to

$$13x + 5y = 1. \tag{2.2.8}$$

Thus $y = \frac{-13x+1}{5} = \frac{-10x-3x+1}{5} = -2x + \frac{-3x+1}{5}$. We must find x and y, both integers, such that Eq. (2.2.8) is satisfied. Guessing in Eq. (2.2.8) is easier than its predecessor above. First, we pick a value of x such that the second term $\frac{-3x+1}{5}$ is an integer. We see $x = 2$ certainly works, which yields $y = -4 + (-1) = -5$ and so $(2, -5)$ is a particular solution of $611x + 235y = 47$. (There are, of course, other choices for x which will work.)

The set of general solutions is, in the case of $x_0 = 2$ and $y_0 = -5$

$$\{(x, y) \mid x = 2 + 235t, y = -5 - 611t \text{ for some } t \in \mathbb{Z}\}.$$

//

Exercise 29. Find a particular solution for an integral linear combination which equals GCD(7, 19) by using the technique of the previous example.

Recall from the previous section that if $a, b \in \mathbb{Z}$, and neither are zero, then a and b are defined to be relatively prime if they have no common positive divisor other than 1. Put more succinctly, $GCD(a, b) = 1$. The following result, which follows from the Euclidean Algorithm Byproduct, is left as Problem 11 of Problem Set 2.2.

Proposition 5. *Let a and b be positive integers. Then a and b are relatively prime if and only if 1 is an integral linear combination of a and b.*

Example 44. Find all positive integers a and b such that

$$a + b = 156 \text{ and } GCD(a, b) = 26.$$

Solution. Since the GCD$(a, b) = 26$, Proposition 3 (page 45) shows that there are relatively prime, positive integers h and k with

$$a = 26h \text{ and } b = 26k.$$

Since $a + b = 156$, we have that

$$h + k = \frac{156}{26} = 6.$$

Therefore, as h and k are relatively prime, we see by examination the only possibility is that one of them is 1 and the other 5. So the only solutions are $a = 26$ and $b = 130$ (or $b = 26$ and $a = 130$). //

Before you look at the next exercise, which you should do in a manner similar to Example 44, think about the following problem:

"Find two integers, whose sum is 84, both of which are divisible by 7 but not by any larger integer."

Without the concepts of GCD and the Euclidean Algorithm Byproduct (Theorem 8), this would be a very difficult problem. However, because we have established machinery (a foundation), you will find the problem solving easy (Exercise 30).

Exercise 30. Find all positive integers a and b such that

$$a + b = 84 \text{ and } GCD(a, b) = 7.$$

Proposition 6. *If a and b are relatively prime and $a|bc$ then $a|c$.*

Proof. If $c = 0$, there is nothing to prove so we may assume that $c \neq 0$. a and b are relatively prime which means that the $GCD(a, b) = 1$. Using the Euclidean Algorithm Byproduct, we may therefore express 1 as a linear combination of a and b. That is, there are integers x and y with

$$1 = xa + yb. \tag{2.2.9}$$

Because $a|bc$, there is an integer k such that

$$ka = bc.$$

Equation (2.2.9), solved for b,

$$b = (1 - xa)/y$$

can be substituted into $ka = bc$ to give:

$$ka = \frac{1 - xa}{y}c \quad \text{or} \quad (ky)a = c - (xc)a.$$

Thus $(ky + xc)a = c$ and so a divides c as $(ky + xc)$ is an integer. ∎

Corollary 3. *If $m|a$ and $n|a$ with $GCD(m, n) = 1$, then $mn|a$.*

Proof. Since $m|a$, $a = mq$ for some $q \in \mathbb{Z}$. Since $n|a$, $n|mq$, and therefore $n|q$ since $GCD(m, n) = 1$. Therefore we have that $q = nt$ for some $t \in \mathbb{Z}$, and so $a = mnt \implies mn|a$. ∎

Exercise 31. Give an example of positive integers a, b and c such that $a|bc$ but a does not divide either b or c. (This exercise shows that the hypotheses that $GCD(a, b) = 1$ cannot be dropped in Proposition 6.)

Problem Set 2.2:

Assume throughout this problems set that neither a nor b are zero.
1. Find the $GCD(46, 69)$ using the Euclidean Algorithm, and use the method of backwards substitution to express it as a linear combination of 46 and 69.
2. Find the $GCD(46, 69)$ by the tableau method and write it as a linear combination of 46 and 69.
3. Use the Tableau Method to write $GCD(136, 62)$ as a linear combination of 136 and 62. Compare your answer to Example 39.

4. Suppose that a and b are integers and that 1 is a linear combination of a and b. Prove that a and b are relatively prime. Compare this result with Theorem 8.

5. If $x_0, y_0 \in \mathbb{Z}$ is a particular solution for $GCD(a, b)$, $t \in \mathbb{Z}$ and if $x = x_0 + bt$ and $y = y_0 - at$ then $GCD(a, b) = xa + yb$. (This shows that every general solution actually gives the $GCD(a, b)$ as an integer linear of a and b.)

6. (a) If a and b are relatively prime, a is odd, and b is even, prove that $(a + b)$ and $(a - b)$ are relatively prime.

   (b) If a and b are both even, prove they are not relatively prime.

   (c) If a and b are both even or both odd and $a \neq b$ prove that $(a + b)$ and $(a - b)$ are never relatively prime.

7. Find all positive integers a and b such that $a + b = 15$ and $GCD(a, b) = 3$.

8. Find all positive integers a and b such that $a + b = 132$ and $GCD(a, b) = 11$.

9. Assume that a and b are both positive integers and x_0 and y_0 are integers with $GCD(a, b) = x_0 a + y_0 b$. Prove that, for any $t \in \mathbb{Z}$, $x = x_0 + bt$ and $y = y_0 - at$ also gives the $GCD(a, b)$ as an integral linear combination of a and b. (That means you must prove that $GCD(a, b) = xa + yb$.)

10. Find two integers whose sum is 192 both of which are divisible by 32 but not both divisible by any larger integer.

11. Prove Proposition 5 and compare with Problem 4.

12. (a) If k is a positive integer, prove that k and $(k + 1)$ are relatively prime.

    (b) Let p be a prime number.

        (1) (experiment) Pick you favorite prime number p and for some examples (such as $n = 1, 2$, up to 15), see when $n + p$ and n are relatively prime.

        (2) (conjecture) What do you think is true? (There should be a pattern in your answer to #1.) Hint: Your conjecture should take the form "if p is a prime number and ????, then $n + p$ and n are relatively prime."

        (3) (proof) Prove your conjecture.

13. Find $m, n \in \mathbb{Z}$ such that $945m + 219n = GCD(945, 219)$ using the Tableau Method.

2.3 Linear Equations with Integer Solutions: Diophantine Equations

Overview In the previous section, one of the key ideas we discussed was the existence of solutions (x, y) to $ax + by = GCD(a, b)$ where x and y are both integers. Suppose we now consider the more general equation in which a, b, and c are integers (with a and b not both zero) such that

$$ax + by = c \qquad (2.3.1)$$

and again look for integer solutions.

The major result of this section is to give necessary and sufficient conditions for Eq. (2.3.1) to have integer solutions. In other words, we will give a condition on the integers a, b, and c so that Eq. (2.3.1) has a solution exactly when this condition is met (Theorem 9). In the case when there is a solution, we will give all of the (infinitely many) solutions. We apply this theory to the "Hundred Fowls" problem from sixth century Chinese mathematical literature. To solve the "Hundred Fowls" problem without the foundation that we will be establishing would be remarkable and it was! Philosophically it is frequently the case that to do problem solving is to build a foundation on existing platforms. Here's its statement—it's solution is at the end of the section.

Hundred Fowls Problem (Burton [1]): If a cock is worth 5 coins, a hen 3 coins, and three chicks together 1 coin, how many cocks, hens and chicks, totaling 100, can be bought for 100 coins?

The Hundred Fowls Problem is a special case of **Diophantine Equation**, mathematics problems whose solution involve only integers. We will only consider the linear case, the subject is quite deep and the machinery needed in general very abstract (and quite beautiful).

Definition. Let a, b, and c be given integers (not all zero). An equation of the form

$$ax + by = c \qquad (2.3.2)$$

is called a LINEAR DIOPHANTINE EQUATION. A DIOPHANTINE SOLUTION to Eq. (2.3.2) is a solution (x, y) in which both x and y are integers. If x, $y > 0$, then it is a SOLUTION IN POSITIVE INTEGERS or a POSITIVE DIOPHANTINE SOLUTION.

It is customary to apply the term *Diophantine equation* to any equation in one or more unknowns that is to be solved in the integers. We will work with only two unknowns. A solution of a Diophantine equation is a pair of integers (x_0, y_0) that, when substituted into the equation, satisfy it; that is, we ask that $ax_0 + by_0 = c$ and that $x_0 \in \mathbb{Z}$ and $y_0 \in \mathbb{Z}$. Thus, the coefficients in a Diophantine equation and the solution must all be integers. Of course, there are infinitely many real solutions (x, y) to (2.3.1)—our question for this chapter is whether there are solutions in which both x and y are integers.

Although the dates of his birth and death aren't known accurately, it is commonly thought that Diophantus was born in about 200 A.D. and died about 284 (both in Alexandria, Egypt). An interesting fact is that there is a crater on the Moon named after him! Curiously enough, the linear equation does not appear in the extant works of Diophantus (the theory required for its solution can be found in Euclid's Elements), possibly because he viewed it as too easy. Most of his problems dealt with finding squares or cubes with certain properties among the integers, rather than the linear equations of Eq. (2.3.2).

Convention When the phrases "solve the Diophantine equation" or "what are the solutions to the Diophantine equation?" are used, it is understood that we are looking for Diophantine solutions; that is, we are searching for integer solutions only.

A given Diophantine equation can have a number of solutions. For example, the equation $3x + 6y = 18$ has solutions $(4, 1)$, $(-6, 6)$, and $(10, -2)$ since

$$3 \cdot 4 + 6 \cdot 1 = 18,$$

$$3(-6) + 6 \cdot 6 = 18,$$

$$3(10) + 6(-2) = 18.$$

(It may also have others; it fact, it does.) By contrast, there is no Diophantine solution to the equation $2x + 10y = 17$, because the left-hand side is an even integer for any choice of the integers x and y, whereas the right-hand side is odd. Of course, there are many real solutions—it's just that none of them have the property that both x and y are integers. Faced with these possibilities, it is reasonable to enquire about the circumstances under which a Diophantine solution is possible and, when a solution does exist, whether we can determine all integer solutions explicitly.

The condition for solvability of the Diophantine equation is easy to obtain, as we'll now prove. We'll first show that the linear Diophantine equation $ax + by = c$ admits a solution if and only if $d \mid c$, where $d = \mathrm{GCD}(a, b)$. Here's the reasoning: if $d = \mathrm{GCD}(a, b)$, then it must divide both a and b. Thus, we know there are integers r and s for which $a = dr$ and $b = ds$.

Now let's assume that there is a Diophantine solution to (2.3.1). If (x_0, y_0) is a Diophantine solution, of $ax + by = c$, then $ax_0 + by_0 = c$. Therefore, from the previous paragraph,

$$c = ax_0 + by_0 = drx_0 + dsy_0 = d(rx_0 + sy_0)$$

which simply says that $d \mid c$ (since $rx_0 + sy_0$ is an integer).

Conversely, assume that $d \mid c$, say $c = dv$ for some $v \in \mathbb{Z}$. We'll prove that this assumption guarantees a solution to (2.3.1). The Euclidean Algorithm Byproduct (Theorem 8) shows that there are integers $\bar{x}$ and $\bar{y}$ which satisfy $d = a\bar{x} + b\bar{y}$. When this relation is multiplied by v we see that

$$c = dv = (a\bar{x} + b\bar{y})v = a(v\bar{x}) + b(v\bar{y}).$$

Hence the Diophantine equation $ax + by = c$ has $x_0 = v\bar{x}$ and $y_0 = v\bar{y}$ as a particular solution. This computation proves the first part of our next theorem. Note that, in order to solve this problem about solutions of a Diophantine equation, we needed to use a lot of the material of the previous section. Now that we know the criteria for Diophantine solutions, we prove the fundamental theorem of linear Diophantine equations by characterizing the solution.

Theorem 9 (Existence and Form of Solutions to Linear Diophantine Equations). *The linear Diophantine equation* $ax + by = c$ *has a Diophantine solution if and only if* c *is divisible by* $d = \mathrm{GCD}(a, b)$. *Furthermore, if* (x_0, y_0) *is any particular solution of this equation, then all other solutions are given by*

$$x' = x_0 + \left(\frac{b}{d}\right)t \qquad\qquad y' = y_0 - \left(\frac{a}{d}\right)t \qquad\qquad (2.3.3)$$

where t *is an arbitrary integer.*

Proof. We have already proven the assertion of the first sentence in the preceding paragraphs. To establish the second statement of the theorem, let us suppose that an integer solution (x_0, y_0) of the given equation is known. If (x', y') is any other integer solution, we now show that (x', y') takes the form of Eq. (2.3.3). Of course,

$$ax_0 + by_0 = c = ax' + by',$$

which is equivalent to

$$a(x' - x_0) = b(y_0 - y').$$

Thanks to Proposition 4 there exist relatively prime integers r and s such that $a = dr$, $b = ds$. Substituting these values into the previous equation and canceling the common factor d, we find that

$$r(x' - x_0) = s(y_0 - y'). \qquad\qquad (2.3.4)$$

Since r is a factor of the left hand side, r must divide the right hand side of (2.3.4) also since $r|s(y_0 - y')$. Since r and s are relatively prime, r cannot divide s and so it must be that case that r divides the other factor (see Proposition 6). Thus $r|(y_0 - y')$, or in other words, $y_0 - y' = rt$ for some integer t. Substituting this into Eq. (2.3.4), we obtain

$$r(x' - x_0) = srt$$

or

$$x' - x_0 = st.$$

Plugging the formula for x' into $ax' + by' = c$, gives y' which leads us to the formulas:

$$x' = x_0 + st = x_0 + \left(\frac{b}{d}\right)t$$

$$y' = y_0 - rt = y_0 - \left(\frac{a}{d}\right)t$$

which are exactly Eq. (2.3.3).

It is easy to see that these values are integers which satisfy the Diophantine equation, regardless of the choice of the integer t because

$$ax' + bx' = a\left[x_0 + \left(\frac{b}{d}\right)t\right] + b\left[y_0 - \left(\frac{a}{d}\right)t\right]$$

$$= (ax_0 + by_0) + \left(\frac{ab}{d} - \frac{ab}{d}\right)t$$

$$= c + 0 \cdot t = c$$
∎

Note, in particular, that Theorem 9 says that there are either no solutions or there is an infinite number of Diophantine solutions to the given equation, one for each integer value of t in Eq. (2.3.3). A geometric way to consider this theorem is described in Problem 5 at the end of this section.

Example 45. We will now use the method of Theorem 9 to solve the linear Diophantine equation

$$16x + 39y = 10. \tag{2.3.5}$$

Note that not only do we have to show that Eq. (2.3.5) has a solution, but we actually have to find all of them!

Solution. To see if there is a Diophantine solution to (2.3.5), we must first, according to Theorem 9, determine if $d = \text{GCD}(39, 16)$ divides 10. A casual glance at 16 and 39 show they have no positive common divisors other than one, so $\text{GCD}(16, 39) = 1$. However, for practice, we will find d using the methods of the last section and then write d as an integral linear combination of 39 and 16. Applying the Euclidean Algorithm, we write

$$39 = 2 \cdot 16 + 7$$
$$16 = 2 \cdot 7 + 2$$
$$7 = 3 \cdot 2 + 1 \tag{2.3.6}$$
$$2 = 2 \cdot 1 + \boxed{0}$$

Therefore, $\text{GCD}(39, 16) = 1$ (16 and 39 are relatively prime) and so there is a solution to Eq. (2.3.5) thanks to Theorem 9.

To solve for a particular solution (which is the next step), we must obtain the integer 1 as a linear combination of $a = 16$ and $b = 39$. As in the last section, we work up from the next to last equation of (2.3.6).

$$1 = 7 - 3 \cdot 2$$
$$= 7 - 3(16 - 2 \cdot 7) = 7 - 3(a - 2 \cdot 7)$$
$$= 7 - 3a + 3(2 \cdot 7) = 7 \cdot 7 - 3a.$$

Since the top equation is $7 = 39 - 2 \cdot 16 = b - 2a$, the above simplifies to

$$1 = 7(b - 2a) - 3a = 7b - 17a \tag{2.3.7}$$

Multiplying Eq. (2.3.7) by 10

$$70b - 170a = 10$$

$$16(-170) + 39(70) = 10.$$

Thus, a particular solution of Eq. (2.3.5) is $x_0 = -170$ and $y_0 = 70$. All solutions are therefore expressed by

$$x = -170 + 39t \text{ and } y = 70 - 16t \tag{2.3.8}$$

for some integer t.   //

There is another way to find a particular solution using an elementary method as follows. We solve the Diophantine equation $16x + 39y = 10$ for x to get the equation $x = \frac{10 - 39y}{16}$. Thus

$$x = -2y + \frac{10 - 7y}{16} \tag{2.3.9}$$

and we are left with the following question. When is $\frac{10 - 7y}{16}$ an integer? We must look for cases when the numerator is a multiple of 16. For example, in the case -480, $10 - 7y_0 = -480$ yields $y_0 = 70$ and $x_0 = -170$. Another solution can be found such as $x_0 = -14$ and $y_0 = 6$.

The above technique can also be used for general solutions. It can best be explained by an example as the following:

$$5x + 22y = 18$$

implies

$$x = \frac{18 - 22y}{5} = 3 - 4y + \frac{3 - 2y}{5}.$$

Since $\frac{3 - 2y}{5} \in Z$, we can write $z = \frac{3 - 2y}{5} = z$. So $2y + 5z = 3$. We now repeat the same argument, solving y for z, namely:

$$y = \frac{3 - 5z}{2} = 1 - 2z + \frac{1 - z}{2}.$$

Let $\frac{1-z}{2} = t$, then $z = 1 - 2t$, $y = (1 - 2z) + t = 1 - 2(1 - 2t) + t = -1 + 5t$ and $x = \frac{18 - 22y}{5} = \frac{18 - 22(-1 + 5t)}{5} = 8 - 22t$.

A third way to find a particular solution is to guess! For example, is the Diophantine problem $7x + 4y = 100$ (which will appear in the solution to the Hundred Fowls Problem as Eq. (2.3.11)), the easy guess of $x = 0$ and $y = 25$ jumps out at us.

Corollary 4. *If a and b are relatively prime and if (x_0, y_0) is a particular solution to the linear Diophantine equation $ax + by = c$, then all solutions are given by the pair (x, y) with*

$$x = x_0 + bt \text{ and } y = y_0 - at$$

for integral values of t.

Here is an easy illustration. The equation $5x + 22y = 18$ has (by guessing) $x_0 = 8$, $y_0 = -1$ as a solution. Since 5 and 22 are relatively prime, Corollary 4 gives a complete solution, $x = 8 + 22t$ and $y = -1 - 5t$ for arbitrary integer, t.

Definition. The solution of a Diophantine equation is a POSITIVE SOLUTION if each of its variables are positive.

Example 46.   What are the positive Diophantine solutions of

$$172x + 20y = 1000. \tag{2.3.10}$$

Solution. In Problem 4 at the end of this section you will show that all solutions of Eq. (2.3.10) are of the form

$$x = 500 + 5t \text{ and } -4250 - 43t \text{ for } t \in \mathbb{Z}.$$

To find solution with $x > 0$ and $y > 0$, t must be chosen to simultaneously satisfy these inequalities. In other words, it must be that both

$$5t + 500 > 0 \text{ and } -43t - 4250 > 0.$$

Therefore, or, what amounts to the same thing,

$$-\frac{4250}{43} = -98\frac{36}{43} > t > -100.$$

Because t must be an integer, we are forced to conclude that $t = -99$. Thus, our Diophantine equation has a unique positive solution, $x = 5$, $y = 7$, corresponding to $t = -99$.   //

Linear indeterminate problems such as those of Diophantus have a long history, occurring as early as the first century in the Chinese mathematical literature. Owing to a lack of algebraic symbolism, they often appeared in the guise of rhetorical puzzles or riddles. The contents of the *Mathematical Classic of Chang Chin-chien* (sixth century) attest to the algebraic abilities of Chinese scholars. This elaborate treatise contains one of the most famous (in the sense of transmission to other societies) problems in indeterminate equations, the problem of the "hundred fowls". The problem quoted at the beginning of this section asks:

Hundred Fowls Problem: (Burton [1]) If a cock is worth 5 coins, a hen 3 coins, and three chicks together 1 coin, how many cocks, hens and chicks, totaling 100, can be bought for 100 coins?

Thanks to the machinery that we have derived in the last and this section, the Hundred Fowls problem is straightforward to solve. Before you read further, think about how you would solve (regardless of whether you live in the twenty-first or sixth century) the Hundred Fowls problem before you had the foundation of Sects. 2.2 and 2.3.

If x equals the number of cocks, y the number of hens, and z the number of chicks then the Hundred Fowls Problem becomes

$$5x + 3y + \frac{1}{3}z = 100 \text{ and } x + y + z = 100.$$

Eliminating one of the unknowns, we are left with a linear Diophantine equation in the other two unknowns. Specifically, because the second equation is $z = 100 - x - y$, we have $5x + 3y + \frac{1}{3}(100 - x - y) = 100$. The Hundred Fowls Problem thus reduces to solving the linear Diophantine equation

$$7x + 4y = 100. \tag{2.3.11}$$

We solve the Hundred Fowls problem using the results of Corollary 4, remembering that in this case $x \geq 0$, $y \geq 0$ and $z \geq 0$ (since you cannot have a negative number of fowls!). We are looking then for a solution in the non-negative integers only.

To carry out our procedure, we need a particular solution to Eq. (2.3.11). Now there is one obvious solution, $(0, 25)$. Corollary 4 shows this equation has the of the form general solution $x = 4t$ and $y = 25 - 7t$ where $t \in \mathbb{Z}$. Because $x + y + z = 100$, $z = 75 + 3t$. Chang himself gave several answers:

$$x = 4 \qquad\qquad y = 18 \qquad\qquad z = 78$$

and

$$x = 8 \qquad\qquad y = 11 \qquad\qquad z = 81$$

and

$$x = 12 \qquad\qquad y = 4 \qquad\qquad z = 84.$$

A little more effort produces all solutions in the positive integers. To find solutions in the positive integers for x and y, t must be chosen to satisfy simultaneously the inequalities

$$4t \geq 0, \qquad 25 - 7t \geq 0, \text{ and also } 75 + 3t \geq 0.$$

The last two of these inequalities are equivalent to $-25 \leq t \leq 3\frac{4}{7}$. Since $4t \geq 0$, t must have a nonnegative integer value, we conclude that $t = 0, 1, 2, 3$, leading precisely to the values that Chang obtained.

Problem Set 2.3:

1. Which of the following Diophantine equations cannot be solved (that is, have no integer solutions)?

    (a) $3x + 15y = 21$
    (b) $30x + 25y = 101$
    (c) $2x + 14y = 91$

2. Determine all solutions in the integers, and positive integers, to the following Diophantine equations:

    (a) $56x + 12y = 60$
    (b) $12x + 18y = 36$
    (c) $101x + 1002y = 1$
    (d) $5x + 22y = 18$

3. Use the method of Theorem 9 to solve the linear Diophantine equation $172x + 20y = 1000$.
4. Prove Corollary 4 by using Theorem 9.
5. Here's a geometric visualization of Theorem 9 which gives all of the solutions to a Diophantine equation. If (2.3.1) has a particular solution, (x_0, y_0), prove that the set of all solutions is the set of all points on the line through (x_0, y_0) with slope $-\frac{a}{b}$ both of whose coordinates are integers.
6. Prove that the Diophantine equation $ax + by + cz = d$ is solvable if and only if $GCD(a, b, c)$ divides d.
7. (a) Is $15x + 12y + 30z = 34$ solvable in integers?
    (b) If $15x + 12y + 30z = k$ is solvable in integers, then what are possible values of k?

2.4 A Brief Introduction to Combinatorics

Overview In previous sections, we have explored some techniques for solving certain linear Diophantine equations. We now turn our attention from *looking* for solutions to that of *counting how many* solutions there are to a given equation. This will require first an introduction to the notions of permutations and combinations followed by their application to Diophantine equations. We do not assume that the reader has a thorough background in the subject, usually called **combinatorics**, although you will have seen the beginnings of combinatorics in high school, in a "math for future teachers" course, or more advanced courses which are dedicated to it.

We then will give some examples but not examine this vast and interesting area as we are only interested in exploring the material for use in number theory. We start with two simple examples.

Example 47. If Richard has 3 sweaters and 4 shirts, how many ways can he pick a pair consisting of a sweater and a shirt?

Solution. If we list the shirts (sh) and sweaters (sw) as in the table below, then each pick of Richard's can be viewed as an arrow joining one of the shirts to one of the sweaters. In the case indicated, it connects shirt 2 with sweater 2.

$$
\begin{array}{ll}
\text{shirts} & \text{sweaters} \\
\text{sh1} & \text{sw1} \\
\text{sh2} \longrightarrow & \text{sw2} \\
\text{sh3} & \text{sw3} \\
\text{sh4} &
\end{array}
$$

Since we can connect a shirt with any of 3 sweaters, it seems clear that the total number of shirt/sweater pairs is $3 + 3 + 3 + 3 = 3(4) = 12$. The answer is just the product of the number of shirts and the number of sweaters.   //

It is important to note that picking a shirt does not influence at all which sweater is chosen. In such cases, we say that the choices are INDEPENDENT. We will be concerned nearly exclusively with counting independent events, but to illustrate any concept (such as independence), it is useful to give an example in which the concept doesn't apply.

Example 48. In a deck of playing cards there are 4 kings and 48 other cards. Jane is going to pick twice. How many ways are there

(A) of getting two kings if Jane replaces the first pick, shuffles the deck and then picks again?
(B) of getting exactly one king if Jane holds the first pick and from the remaining 51 cards picks again?

Solution.

(A) Jane has 4 choices of a king the first time and 4 the second time so the answer is 16. (The events of picking are independent—she's doing the same thing twice).
(B) In this case, we can't just multiply because what happens on her first pick changes the number of choices. There are 4 choices of a king the first time she picks. We now must divide the possibilities into two cases. The first is that she

is successful on the first pick (she gets one of the 4 kings). She must then pick
something other than a king and there are 48 ways to do that. Thus

> Number of ways of picking a king followed by another card which is not a
> king = 4(48) = 192.

On the other hand, if she gets a non-king first (48 possibilities) and then a
king (4 possibilities), there are (48)4 = 192 ways. Thus there are a total of
192 + 192 = 384 choices.   //

What we see in Case B is that we must be careful as to whether the events are
independent in Jane's experiment. To be more precise, the event of picking a king
and then the second event of another king from the original deck are independent
(Part A). The event of picking only one king involves dependent events. This is the
difference between picking with replacement versus without replacement. Note that
the answer to Part B is NOT 4(48), which would be the answer to "first pick a king
and then pick a non-king without replacement of the first card".

Deciding whether two events are independent can be tricky. Culture and taste
come into the equation, as we can see in the next example.

Example 49. A restaurant serves chicken, steak, tofu, beef, mashed potatoes,
French fries, baked potatoes and green beans. How many different ways are there to
select two items for your meal.

Proof (Solution for Discussion). Think about and share ideas with friends for what
a solution even means! Would you want 2 orders of mashed potatoes as your
entire meal? (In the language of Example 48, this is the question of picking "with
replacement" or "without replacement".)

Here are other "practical questions": Would you want an order of chicken and an
order of steak? Do you eat tofu? Are you a vegetarian?

There is no correct answer to this problem until the exact criteria for the meal are
specified.   ∎

It should be apparent the notation for these types of problems could become very
cumbersome if we must make many choices. To motivate our notation, suppose
you have 4 red balls, 3 white ones, 2 green ones and a yellow one. Then there are
$4 \cdot 3 \cdot 2 \cdot 1 = 24$ ways of picking a red, white, green and yellow ball in that order. This
kind of computation has a very specific formula which leads us to make a definition
which helps with writing formulas succinctly.

Definition. If n is a positive integer, n FACTORIAL, written as $n!$, is

$$n! = n(n-1)\cdots 3 \cdot 2 \cdot 1.$$

We also define 0! as 1. Note that $3! = 3 \cdot 2 \cdot 1 = 6$ and $4! = 4 \cdot 3 \cdot 2 \cdot 1 = 24$. In fact,
$4! = 4 \cdot 3!$.

Exercise 32. Prove that for n an integer at least 1

$$n! = n \cdot (n-1)!$$

Example 50. As in Example 48, Jane again wants to pick two cards but this time she doesn't replace the first card into the deck. Assuming the order of kings matters (so getting the king of hearts and then getting the king of spades is different than getting the king of spades and then the king of hearts), how many ways are there of getting 2 kings?

Solution. She can find 4 opportunities for the first king and 3 for the second. Thus there are $12 = 4 \cdot 3$ ways.   //

Note that the answer to Example 50 can be written as

$$4 \cdot 3 = \frac{4!}{2!} = \frac{n!}{(n-r)!}, \text{ where } n = 4, \text{ and } r = 2.$$

This kind of problem (how many ways are there of picking r objects in order without replacement from a total of n objects?) comes up so often in problem solving that its solution has a special name, permutation, and symbol $_nP_r$.

Definition. Let n and r be non-negative integers, with $n \geq r$. A PERMUTATION of n objects taken r at a time is

$$_nP_r = \frac{n!}{(n-r)!}.$$

Suppose now that we consider a different problem, one in which the objects we choose are not all different.

Example 51. A n-CODE WORD is a string of n letters. (It need not make sense in any language). How many 5-code words can be made from the letters of the set $\{T, U, V, W, X\}$ if

(A) the 5 letters are distinct?
(B) 3 of the letters are W and the others are T and U?

Solution.

(A) Since the letters are all different, there are 5 choices for the first, 4 for the second, etc. Thus the answer is

$$5 \cdot 4 \cdot 3 \cdot 2 \cdot 1 = 5!$$

which is 120.
(B) To solve this part of the problem, let's first make the W's "look different" by adding subscripts to them. The possibilities are then

$$T, U, W_1, W_2, W_3.$$

Since they are all different now, we know from Part A that the answer is 5!. However, since the W's are all the same the following three different words are actually the same $UW_2TW_3W_1$, $UW_1T W_2W_3$, and $UW_2TW_1W_3$. Each of them is the code word *UWTWW* of the original problem. In fact, if U is in the first place and T is in the third then there are 6 code words which appear in Part A but would be identical in Part B. Thus the answer to Part B is $\frac{5!}{6}$ or $\frac{5!}{3!}$, since $6 = 3!$.   //

Note that to avoid the cultural aspects of what a "word" is, we defined it carefully in the problem. Compare the statement of Example 51 with that of Example 49.

Exercise 33. How many 5-code words from Part A of the previous example become *UWTWW* in Part B?

From this discussion, we see that if there are n objects such that k of them are identical and the other $(n - k)$ distinct and differ from the first k objects, then the number of code words is $n!/(n - k)!$ which is exactly the number of permutations of n things taken k at a time, $_nP_r$.

Let's move on to a more complex situation in which the same logic holds and introduce the idea and notation of "combinations," in contrast to permutations. The key difference between the two is whether the order of the groupings matters.

Suppose we have n different people and we wish to pick r of them ($r \leq n$) for a committee. This "committee problem" gives a good insight into the difference between a combination and a permutation. In a permutation, the order of choice matters (think about picking a president, a vice president, and a treasurer of an organization). For a committee of three, it doesn't matter in which order you pick the committee members—it's still the same committee.

Since the people of our committee are different, it might seem like the answer should be $n(n - 1) \cdots (n - r + 1)$. However, this is a committee so the *order of choosing the people doesn't matter*. While people are different, what matters is whether someone is chosen for the committee or not. Two people are then "identical" for the problem we are considering if they are both on the committee or both not on the committee. Thus if we count permutation, we have counted the same committee $r!$ times. (This is just like the code word problem where we counted the same code word 3! times.) The answer to the committee problem is obtained by dividing by the number of times we have counted the same committee.

$$\text{number of committees of } r \text{ members} \over \text{of a population of } n} = \frac{n(n - 1) \cdots (n - r + 1)}{r!} \qquad (2.4.1)$$

This expression comes up so often in counting that it has its own name and symbol, $_nC_r$ or $\binom{n}{r}$. You should read $_nC_r$ or $\binom{n}{r}$ as either "n choose r" or "the number of combinations of n things taken r at a time". $_nC_r$ represents the number of ways of picking r objects (from n) *without regard to order*.

Definition. If $n \geq 0$ and $r \geq 0$ are integers with $n \geq r$, then the COMBINATIONS OF n THINGS TAKEN r AT A TIME is

$$_nC_r = \binom{n}{r} = \frac{n!}{r!(n - r)!}. \qquad (2.4.2)$$

Note that $\binom{n}{r}$ is exactly the expression in Eq. (2.4.1) (and we will use $\binom{n}{r}$ or $_nC_r$ interchangeably).

Note $\binom{n}{r}$ is also the number of subsets of an n-element set that contains r elements.

Example 52. By (2.4.2), we see that the number of subsets containing 3 elements that can be formed from the set $\{a,b,c,d,e\}$ is $\binom{5}{3} = \frac{5!}{3!\,2!} = \frac{5\cdot4\cdot3!}{3!\,2!} = \frac{5\cdot4}{2\cdot1} = 10$.

The ten subsets in question are $\{a,b,c\}$, $\{a,b,d\}$, $\{a,b,e\}$, $\{a,c,d\}$, $\{a,c,e\}$, $\{a,d,e\}$, $\{b,c,d\}$, $\{b,c,e\}$, $\{b,d,e\}$, $\{c,d,e\}$

Example 53. How many ways can a committee of four women be chosen from six women?

Solution. It is equivalent to find the number of subsets containing 4 elements that can be formed from a 6 element set. Therefore, the answer is $\binom{6}{4} = \frac{6!}{4!\,2!} = \frac{6\cdot5}{2\cdot1} = 15$.   //

Problem Set 2.4:

1. Find the value of each of the following.
   (a) $\binom{100}{80}$ (b) $\binom{15}{2}$ (c) $\binom{11}{5}$

2. If Jeff has 3 sweaters, 5 shirts and 3 pants, then how many ways can Jeff pick an outfit (one sweater, one shirt, and one pair of pants) if

   (a) there is no restriction on his choice.
   (b) there is one red sweater, one red shirt, and one red pair of pants and all the other cloths are different colors (and not red) and Jeff won't wear an outfit with two red items?
   (c) If there are 3 red shirts in Part B instead of one, answer the same question.

3. If Maria is playing with a standard deck of cards, how many ways can she

   (a) pick two cards with replacement each of which is less than 6? (An ace should be counted as low in this exercise.)
   (b) pick two cards without replacement each of which is less than 6?

4. (a) How many 6-code words can be made from letters in the English alphabet? (Do not do all of the arithmetic to get a specific number; a formula is fine.)
   (b) Answer Part (a) if the first choice must be a vowel (the letter y is not a vowel for this exercise.)
   (c) Answer Part (a) if there must be two vowels in each of the 6-code words.

5. Are the next equations true for positive integers n and m? In other words, either prove or give a counterexample to each of the following assertions:

   (a) $(n^2)! = n!n!$
   (b) $(n+m)! = n! + m!$
   (c) $\frac{n!}{m!} = n(n-1)\cdots(n-m)$ if $n \geq m$

6. Show that (2.4.2) and (2.4.1) are the same expression.
7. Prove that $\binom{n}{r} = \binom{n}{n-r}$.
8. Prove that $\binom{n}{r} = \binom{n-1}{r-1} + \binom{n-1}{r}$ by using (2.4.2). Can you prove it without doing anything formal? (see Problem 7) This identity was proven by B. Pascal (1623–1662). Therefore it is called the Pascal's identity.
9. A student must answer six of the ten questions on a final exam. How many different ways can he complete the exam.
10. $\binom{52}{5} = 2,598960$. If $\binom{52}{k} = 2,598960$ and $k \neq 5$, then k is equal to what integer?
11. A committee of three—Chair, Secretary, and Treasurer—is to be elected by a club with 14 members. If every member is eligible to stand for each position, how many different committees are possible.
12. What is the sum of $\binom{8}{0} + \binom{8}{1} + \binom{8}{2} + \cdots + \binom{8}{8}$?

It's interesting to see that Problem 7 has to be true without doing anything formal (although you should). To form a committee of r people from n, we simultaneously form a committee of $(n - r)$ also (the ones who were not chosen). Thus the number of ways of choosing r objects from n should be the same as the number of ways of choosing $(n - r)$ from n. Problem 7 is a formal statement of this heuristic argument.

2.5 Linear Diophantine Equations and Counting

We now turn our attention on counting how many solutions there are to a given linear Diophantine equation with unit coefficients. We begin with a simple example.

Example 54. How many positive integer solutions are there for the equation $x + y + z = 6$?

Solution. The number of solutions is so small that they may be listed easily. In fact, the possibilities for (x, y, z) are:

$$(4, 1, 1), (3, 2, 1), (3, 1, 2), (2, 3, 1), (2, 2, 2), (2, 1, 3), (1, 4, 1), (1, 3, 2),$$

$$(1, 2, 3), (1, 1, 4) \qquad //$$

There is an easy method for finding the number of positive solutions to such equations, suggested by the following equation:

$$1 + 1 + 1 + 1 + 1 + 1 = 6.$$

To select a solution (x, y, z), we need only chose two of the plus signs to break the sum up into three parts. For instance, $1 \oplus 1 + 1 + 1 \oplus 1 + 1 = 6$ corresponds to $1 + 3 + 2 = 6$ or the solution $(1, 3, 2)$. This can be done in $\binom{5}{2} = 10$ ways. Generalizing this example we have the following theorem.

Theorem 10. *The equation* $x_1 + x_2 + x_3 + \cdots + x_n = m$ *has exactly* $\binom{m-1}{n-1}$ *positive integer solutions.*

Proof. Each choice of $n - 1$ plus signs from the $m - 1$ plus signs in the sum corresponds to a positive integer solution. ∎

Example 55. How many positive integer solutions are there for the equation $x_1 + x_2 + x_3 + x_4 = 16$?

Solution. Here, $m = 16$ and $n = 4$, so there are $\binom{15}{3} = 455$ positive integer solutions. //

Example 56. In an undergraduate dormitory there are several freshmen, sophomores, juniors, and seniors. In how many ways can a team of 12 students be chosen such that the team contains at least one student of each year?

Solution. Let x_1, x_2, x_3, and x_4 be the number of students that are freshmen, sophomores, juniors, and seniors in a team respectively. Then $x_1 + x_2 + x_3 + x_4 = 12$. $x_1 > 0$, $x_2 > 0$, $x_3 > 0$, $x_4 > 0$. Thus by Theorem 10 with $m = 12$ and $n = 4$, we obtain

$$\binom{12-1}{4-1} = \binom{11}{3} = \frac{11!}{3!\,8!} = \frac{11 \cdot 10 \cdot 9}{3 \cdot 2 \cdot 1} = 165$$

//

To find the number of solutions in non-negative integers to our equation $x_1 + x_2 + \cdots + x_n = m$, we have the following theorem.

Theorem 11. *The equation* $x_1 + x_2 + \cdots + x_n = m$ *has exactly* $\binom{m+n-1}{n-1}$ *non-negative integer solutions.*

Proof. Let $y_1 = x_1 + 1$, $y_2 = x_2 + 1$, $y_3 = x_3 + 1, \cdots, y_n = x_n + 1$. Thus, $x_1 = y_1 - 1$, $x_2 = y_2 - 1, \cdots, x_n = y_n - 1$. Substituting into our equation gives

$$(y_1 - 1) + (y_2 - 1) + \cdots + (y_n - 1) = m \quad \text{i.e.} \quad y_1 + y_2 + \cdots + y_n = m + n$$

which has $\binom{m+n-1}{n-1}$ positive integer solutions. Therefore, $x_1 + x_2 + \cdots + x_n = m$ has exactly $\binom{m+n-1}{n-1}$ non-negative integer solutions. ∎

Example 57. A bakery makes 5 different types of donuts. How many different assortments of one dozen donuts can be purchased?

Solution. Let x_i ($i = 1, 2, 3, 4, 5$) denote the 5 different types of donuts. The number of different assortments of one dozen donuts can be rephrased as the number of possible non-negative integer solutions to the following equation.

$$x_1 + x_2 + \cdots + x_5 = 12, \quad x_i \geq 0.$$

Which, by Theorem 11, is $\binom{12+5-1}{5-1} = 1,820$. //

Example 58. Nine professors are eating lunch together at a restaurant that offers four different desserts. How many different selections of desserts can be ordered by the group?

Solution. Let x_i $(i = 1, 2, 3, 4)$ denote the different deserts offered by the restaurant. Thus, the number of selections of desserts that can be ordered is the number of non-negative integer solutions to the following equation.

$$x_1 + x_2 + x_3 + x_4 = 9.$$

Thus, by Theorem 11, the answer is $\binom{9+4-1}{4-1} = 220$. //

Note If we wanted to allow the possibility that some professors might order no dessert, then there are 5 types of individual dessert orders.

Finally, suppose we are interested in restricted solutions to an equation.

Example 59. How many integer solutions are there to the equation $x + y + z + t = 19$ with $x > 0$, $y \geq 7, z, t \geq 0$?

Solution. Let $u_1 = x$, $u_2 = y - 6$, $u_3 = z + 1$ and $u_4 = t + 1$. Then the original equation is equivalent to

$$u_1 + u_2 + 6 + u_3 - 1 + u_4 - 1 = 19, \quad \text{i.e.,} \quad u_1 + u_2 + u_3 + u_4 = 15.$$

Thus, by Theorem 10, we have $\binom{15-1}{4-1} = 364$ solutions. //

Note We could have used Theorem 11 by setting $u_1 = x - 1$, $u_2 = y - 7$, $u_3 = z$, and $u_4 = t$.

Example 60. In how many ways can a team of 12 be chosen such that it has at least one freshman, at least one sophomore, at least two juniors, and at least three seniors?

Solution. Let x_1, x_2, x_3, x_4 be the number of students from each year. In this case, we have

$$x_1 + x_2 + x_3 + x_4 = 12, \quad x_1 \geq 1, x_2 \geq 1, x_3 \geq 2, x_4 \geq 3.$$

Let $u_1 = x_1$, $u_2 = x_2$, $u_3 = x_3 - 1$, and $u_4 = x_4 - 2$. Now,

$$u_1 + u_2 + u_3 + 1 + u_4 + 2 = 12 \Rightarrow u_1 + u_2 + u_3 + u_4 = 9 \quad u_1, u_2, u_3, u_4 \geq 1$$

By Theorem 10, we have $n = 4$ and $m = 9$ and the answer is $\binom{9-1}{4-1} = 56$. //

Problem Set 2.5:

Find the number of integer solutions to the following equations satisfying the given restraints.

1. $x + y + z = 8$, $x \geq 1, y \geq 0, z \geq 0$.

2. $x + y + z + u = 15, x \geq 1, y \geq 1, z \geq 0, u \geq 2$.
3. $x + y + z = 10, x \geq 0, y \geq 2, z > 1$.
4. In how many ways can 20 pieces of fruit be selected from a basket of bananas, cherries, and apples with at least 1 piece of each type of fruit being selected? How many different baskets of fruit can be prepared?
5. The University of Nevada, Las Vegas represents a large group of freshmen, sophomores, juniors, seniors, and graduate students. Find the number of ways of forming a team of 20 UNLV students such that

   (a) at least two members of the team are seniors and there is at least one member from each of the other groups.
   (b) at least two of the team are seniors and the team contains atleast one graduate student.

6. A group of nine children have set out to find 72 tennis balls around a group of trees next to John's home to put in their own box. In how many ways can the balls be distributed in the boxes?
7. A collection has a large set of roses which are red, yellow, pink, and orange. How many different combinations (counting color) can be obtained by choosing ten roses?
8. A restaurant offers five different juice drinks at a price of $2 each. How many selections of juice can a customer with $20 order?
9. What is the number of non-negative integer solutions of $x + y + z + u \leq 14$? (Hint: $x + y + z + u \leq 14$ if and only if $x + y + z + u + t = 14$ with $t \geq 0$.)

Chapter 3
Equivalence Classes with Applications to Clock Arithmetic and Fractions

Introduction We will first define a quite unifying concept: what it means for two objects to be equivalent (rather than equal), and apply it in the next section to modular (or clock) arithmetic and in the following one to establish a rigorous definition of fractions. The notion of equivalence is pervasive in mathematics and will be a part of many courses you will take in the future, including abstract algebra and geometry. The last section shows that there are fascinating applications from number theory.

3.1 Equivalence Relations and Equivalence Classes

Overview While modular arithmetic (the arithmetic of a clock is an example) is quite intuitive, fractions are not. The arithmetic of fractions is generally difficult for children to learn, hard to truly understand conceptually, extremely difficult to define carefully, and applicable. Modular arithmetic is interesting and widespread. Mathematics requires clear, concise definitions. Before we can ask whether 7, 1, or 5 are prime numbers, for example, we must have a precise definition of the concept "prime number". Why is 1 not a prime? Because the very definition of a prime number excludes 1 being a prime number (and all mathematicians have adopted that definition). Occasionally there are definitions which differ in various books, but that occurs rarely. Because definitions are so important in mathematics, going back to the definition of a concept may be necessary in order to truly understand what a question is asking or why a proof is constructed the way that it is. Thus *"looking at definitions" can be thought of as a problem solving strategy.*

Some concepts are easily defined (e.g. a square is a rectangle with all sides the same length) and others are more difficult. The next two concepts of the text (modular arithmetic and fractions) both require the notion of an equivalence relation on a set.

© Springer International Publishing Switzerland 2015
R.S. Millman et al., *Problems and Proofs in Numbers and Algebra*,
DOI 10.1007/978-3-319-14427-6_3

What is a succinct way of differentiating between equality and equivalence? Consider a room full of people. If we are interested in determining a fine basketball team of five of these people then the quality of the basketball skills matters greatly— that is, each person will add something different to the team. No person is "equal" to another in any reasonable sense. On the other hand, if we are interested in performing interviews which requires siblings then we don't really care about individuals, only whether two people are related as brothers, sisters, or brother/sister. While the brothers Philip and Steven are different, they are the same for our experiment. We would say then that Philip and Steven are "equivalent under the sibling relationship" but not equal.

We will now formally define the products of sets, relations on a set and equivalence relations. These notions are the heart of much of mathematics and certainly are the backbone of the next two sections. You will use them often in future courses.

Definition. If A and B are sets, the CARTESIAN or CROSS PRODUCT of A and B is

$$A \times B = \{(a, b) \mid a \in A \text{ and } b \in B\}.$$

$A \times B$ consists of pairs of elements of A and B (in that order).

Example 61. $\mathbb{R}^2$ is defined in most text as $\mathbb{R}^2 = \{(x, y) \mid x \in \mathbb{R}, \ y \in \mathbb{R}\}$ and so $\mathbb{R}^2 = \mathbb{R} \times \mathbb{R}$ is the plane, as usual.

Informally, a relation on a set A is a way of deciding whether, given $a_1, a_2 \in A$, a_1 is related (as in the sibling example above) to a_2. Intuitively, a relation on a set is something where we can determine whether or not two elements are related (true) or not (false).

Definition. Let A be a set. R is a RELATION on A if R is a function from $A \times A$ to the set of two elements $\{T, F\}$. If $R(a_1, a_2) = T$, we say that a_1 and a_2 are RELATED BY R. If $R(a_1, a_2) = F$, then a_1 and a_2 are NOT RELATED BY R.

When you read the next paragraph and the examples, keep in your mind the "sibling relation" for the set of people A, $T = $ true and $F = $ false. In that context $R(a_1, a_2) = T$ means that person a_1 and person a_2 are "in truth" brothers or sisters whereas $R(a_1, a_2) = F$ means that person a_1 and person a_2 are not siblings.

Going back to the general case of R being a relation of a set A, we are thinking that, if $a_1, a_2 \in A$ then $R(a_1, a_2) = T$ (for true) if a_1 is related to a_2 and $R(a_1, a_2) = F$ (for false) if a_1 is not related to a_2. The key here is that the defining property of R is either true or false but not both. Since we'll usually be dealing with relations defined by equations, it will be easy to decide whether or not the numbers are related. It is usually easier to write $a_1 R a_2$ to mean that a_1 is related by R to a_2 or $R(a_1, a_2) = T$. We'll use both notations as is customary in most books.

Definition. Let $A = \mathbb{R}^1$ and $a_1, a_2 \in A$. Then we define the ABSOLUTE VALUE RELATION to be $a_1 R a_2 = T$ exactly when $|a_1| = |a_2|$ (so $a_1 R a_2 = F$ when $|a_1| \neq |a_2|$). Thus a_1 and a_2 are related under the absolute value relation if and only if $|a_1| = |a_2|$.

Definition. Let $A = \mathbb{R}^2$. If $a_1 = (x_1, y_1)$ and $a_2 = (x_2, y_2)$ then we define the NORM RELATION as $a_1 R a_2 = T$ if $x_1^2 + y_1^2 = x_2^2 + y_2^2$ and $R(a_1, a_2) = F$ if $x_1^2 + y_1^2 \neq x_2^2 + y_2^2$.

In the first definition above, 1 is absolute value related to -1 and to itself, but to no other real number. In the second definition above, note that $\left(\frac{1}{\sqrt{2}}, \frac{1}{\sqrt{2}}\right)$ is norm related to $\left(-\frac{1}{2}, \frac{\sqrt{3}}{2}\right)$ since the sum of the squares of their coordinates is 1 in either case. If fact, (x_1, y_1) and (x_2, y_2) are norm related exactly when they both lie on the same circle about the origin of the same radius provided that neither is $(0,0)$. The point $(0,0)$ has no other element of $\mathbb{R}^2$ "norm related" to it, whereas all other points do have (many) other points which are related to them.

There will be two relations that will be especially important to us: relating integers (which we will define here and use in modular arithmetic in Sect. 3.2), and relating fractions in Sect. 3.3.

Suppose that we are interested in a relation based on whether two integers a_1 and a_2 give the same remainder on division by 3. This would mean that 2 is related to 5, 11, and 20 (among others) but 2 is not related to 4. Can we give the relation in other terms? Yes; to say that a_1 and a_2 have the same remainder on division by 3 is to say that $a_1 - a_2$ is divisible by 3. For example, in this relation 2 is related to 20 because both have remainder 2 on division by 3. We are thus led to the formal definition of relation modulo 3 on the set of integers; $a_1 R a_2$ if there is an integer k such that $a_1 - a_2 = 3k$. More generally,

Definition. Let n be a positive integer $(n \geq 1)$. We say that a_1 and a_2, both integers, are CONGRUENT MOD n (or just simply CONGRUENT) if $n | (a_1 - a_2)$ or equivalently, there is an integer k such that

$$a_1 - a_2 = kn. \tag{3.1.1}$$

We write $a_1 \equiv a_2 \pmod{n}$ if a_1 and a_2 are congruent modulo n and may say that a_1 is congruent mod n to a_2 as a shorthand.

Example 62. Show each of the following.

A. $5 \equiv 17 \pmod{6}$.
B. $-1 \not\equiv 23 \pmod{21}$.
C. $-15 \equiv -3 \pmod{6}$.

Solution. Each of these amounts to using the definition which is contained in Eq. (3.1.1). That is, given a_1, a_2 and n, is there a $k \in \mathbb{Z}$ which solves (3.1.1)?

A. $5 - 17 = 12 = 2(6)$ so $k = 2$ works in Eq. (3.1.1), and, yes, $5 \equiv 17 \pmod{6}$
B. $-1 - 23 = -24$ but there is no integer, k, with $-24 = 21k$, so $-1 \not\equiv 23$ (mod 21).
C. $-15 - (-3) = -12 = (-2)6$, so $-15 \equiv -3 \pmod{16}$. //

Example 63.

A. Show that two integers are congruent modulo 2 if and only if both are even or both are odd.
B. Show that every integer is congruent to 0,1 or 2 modulo 3.

Solution.

A. Suppose that a_1 and a_2 are given and both are even. (The case where both a_1 and a_2 are odd is Problem 1 of Problem Set 3.1.) Then there are two integers k_1 and k_2 such that

$$a_1 = 2k_1 \text{ and } a_2 = 2k_2.$$

Thus $a_1 - a_2 = 2(k_1 - k_2)$. Because we may let $k = k_1 - k_2$ in Eq. (3.1.1) we have that $a_1 \equiv a_2 \pmod 2$.

Now suppose a_1 is even and a_2 is odd. We must show that a_1 and a_2 are not congruent modulo 2. Once again, there are $k_1, k_2 \in \mathbb{Z}$ so that $a_1 = 2k_1$ and $a_2 = 2k_2 + 1$ so that

$$a_1 - a_2 = 2(k_1 - k_2) - 1.$$

Thus $a_1 - a_2$ is not divisible by 2 (since -1 isn't).

B. Let $a \in \mathbb{Z}$. We will show that a is congruent to 0, 1, or 2 modulo 3 by showing that $a \equiv r \pmod 3$ is the remainder of a on division by 3. The Division Algorithm (Eq. (∗) of Sect. 1.1) shows that there are integers q and r with $0 \le r < 3$ such that

$$a = q3 + r$$

or $a - r = 3q$. Since $3q$ is divisible by 3, so is $a - r$ and so $a \sim_3 r$. Because $0 \le r < 3, r = 0, 1,$ or 2.   //

The example of relations modulo n is important in number theory and gives the basis for clock arithmetic that is covered in elementary school. To say that a_1 is congruent to a_2 modulo n means that a_1 and a_2 differ by a multiple of n. This is the same as saying that a_1 and a_2 have the same remainder on division by n. For the clock interpretation, we'll use $n = 12$. Informally, we identify (when not in Europe or on Military time) 13 o'clock with 1, 18 o'clock with 6 and 0 o'clock (midnight) with 12 (noon).

Here's a problem in clock arithmetic. What is 3 h after 10 o'clock (on a 12 h clock)? It is 1 o'clock, of course, but think what $10 + 3$ is. $10 + 3 = 13$ is not equal to 1, but it is related to 1. This easy computation hints that there is a similarity between numbers that are equivalent modulo 12 and the arithmetic of clocks as we'll see in Proposition 8 .

We shall now define certain properties of relations which are appealing (some properties require that an element be related to itself, others are symmetric, etc.).

Definition. Let R be a relation on a set, A.

R is a REFLEXIVE RELATION if aRa for all $a \in A$.

R is a SYMMETRIC RELATION if aRb implies that bRa for all $a, b \in A$.

R is a TRANSITIVE RELATION if aRb and bRc means that aRc for all a, b, and $c \in A$.

Definition. Let $P = (x_0, y_0)$ be an element of $\mathbb{R}^2$. The NORM of P is

$$|P| = \sqrt{x_0^2 + y_0^2}.$$

Example 64. If A is the set of all women in your town, and R is the relation "is the sister of," is R a symmetric, reflexive, or transitive relation on A?

Solution. R is certainly symmetric, but most would say that Sandy is not the sister of Sandy so R is not reflexive. Whether R is transitive depends on your definition of "sister". If it includes half-sister then it is possible for (Sandy)R(Cathy) and (Cathy)R(Judy) but for Sandy to not be related to Judy. (Why?) Furthermore, (Cathy)R(Sandy) and (Sandy)R(Cathy) but is it true that (Cathy)R(Cathy)?   //

Exercise 34. Show that the absolute value and norm relations are symmetric, reflexive and transitive.

Definition. We define the GREATER THAN OR EQUAL RELATION on $\mathbb{Z}$ to be aRb if $a \geq b$.

Exercise 35. Prove or find a counterexample to each of the following statements.

A. The Greater than or equal to relation is symmetric.
B. The Greater than or equal to relation is reflexive.
C. The Greater than or equal to relation is transitive.

Definition. A relation R is an EQUIVALENCE RELATION if R is reflexive, symmetric, and transitive. If R is an equivalence relation, we say that a is EQUIVALENT (UNDER R) to b if aRb or, more frequently, an equivalence relation is written $\sim$ or $a \sim b$.

An easy example of an equivalence relation on the set of all finite sets is to say that $S_1 R S_2$ if S_1 and S_2 have the same number of elements. While it is easy for us as adults, that relation is quite abstract for young children. Watching a 3, 4, 5, or 6 year old struggle with why three bananas are "equal" to three apples is fascinating. In our language, the sets are not equal, they are just equivalent under the relation "same number of elements". As adults, we are so used to this abstraction that we don't even think about it. On the other hand, for people new to the idea of sets and the number of elements in them (e.g. children), it is difficult. While we are most certainly not advocating teaching set theory and equivalence relations to young children, our point is that the concepts which, when first learned, were abstract and difficult become automatic with practice and understanding.

Exercise 36. Is the "absolute value relation" from the definition at the beginning of this section an equivalence relation? Why?

Proposition 7. *For any integer $n > 0$, congruent modulo n ($\equiv_n$) on $\mathbb{Z}$ is an equivalence relation.*

Proof. Let $n > 0$. To show that congruent modulo n($\equiv_n$) is an equivalence relation, we must prove the $\equiv_n$ is reflexive, symmetric, and transitive.

Reflexive: If $a \in \mathbb{Z}$, we need to show that $a \equiv_n a$. Looking at Eq. (3.1.2), that is the same as showing that there is a $k \in \mathbb{Z}$ such that $a - a = kn$. Setting $k = 0$ shows that there is such a k and so $a \equiv_n a$ for any $a \in \mathbb{Z}$.

Symmetric: Let $a, b \in \mathbb{Z}$. Assume that $a \equiv_n b$. We will show that $b \equiv_n a$. If $a \equiv_n b$, then there is an integer k with $a - b = kn$ (3.1.1) so that $(b - a) = (-k)n$. Because $(-k)$ is also an integer, $b \equiv_n a$ and so $\equiv_n$ is symmetric.

Transitive: Let a, b, and $c \in \mathbb{Z}$. To show that congruent modulo n is transitive, we assume that $a \equiv_n b$ and $b \equiv_n c$ and show that a and c are related modulo n. We know from the definition of $\equiv_n$ that there are integers k_1 and k_2 with

$$a - b = k_1 n \text{ and } b - c = k_2 n.$$

Adding these two equations gives $a - c = (k_1 + k_2)n$. Since $k_1 + k_2 \in \mathbb{Z}$, $a \equiv_n c$ and $\equiv_n$ is transitive.   ∎

Language. Because congruent modulo n is actually an equivalence relation, we will say that "a is equivalent to b modulo n" or "a is equivalent to b mod n" to mean the same thing as "a is congruent to b modulo n" and use the same symbol $a \equiv_n b$ for both.

The usual usage in mathematics is "a is equivalent to b modulo n" but we wanted to point out that if a relation has the name "equivalence" in its title that doesn't mean it automatically is an equivalence relation. We still must prove that it is reflexive, symmetric and transitive.

Definition. If R is a relation on the set A and $a \in A$ then the RELATION CLASS of a with respect to R, written $[a]_R$, is

$$[a]_R = \{b \in A \mid aRb\}.$$

If R is the relation "congruent modulo n" and $a \in \mathbb{Z}$ we will write $[a]_n$ to be the CONGRUENT CLASS of a modulo n. If the relation being used is obvious, we may write $[a]$ instead of $[a]_n$ where n is a positive integer.

We have already done examples which involve the congruent class. In particular, Example 62A shows that 5 is in the congruent class of 17 modulo 6 (or, in symbols, $5 \in [17]_6$ and $17 \in [5]_6$). Example 62B is the statement that $-1 \notin [23]_{21}$. On the other hand, Example 63A gives a complete description of the congruent class with respect to "related modulo 2." There are two different congruent classes, one is the set of even integers and the other is the set of odd integers. In symbols

$$\mathbb{Z} = \text{odd integers} \cup \text{even integers} \tag{3.1.2}$$

or

$$Z = [0]_2 \cup [1]_2 = [8]_2 \cup [-7]_2. \tag{3.1.3}$$

The equality (3.1.3) is a bit subtle. We are *not* asserting that $0 = 8$ or that $1 = -7$, only that $[0]_2 = [8]_2 = \{\text{even integers}\}$ and $[1]_2 = [-7]_2 = \{\text{odd integers}\}$. These last statements become $0 \sim_2 8$ and $1 \sim_2 -7$. Notice that there are two different congruent classes; they are either disjoint or equal and their union is $\mathbb{Z}$.

Example 65. Let R be the norm relation defined earlier. If $(x_0, y_0) \in \mathbb{R}^2$ and $r = \sqrt{x_0^2 + y_0^2}$ then the relation class of (x_0, y_0) with respect to $\sim$ is a circle of radius r about the origin unless $r = 0$, in which case $[(0,0)]_R = \{(0,0)\}$.

Solution. Let $(x_0, y_0) \in^2$. Then

$$[(x_0, y_0)]_\sim = \{(x, y) \in \mathbb{R}^2 \mid x^2 + y^2 = x_0^2 + y_0^2\}.$$

Thus $[(x_0, y_0)]_R = \{(x, y) \in \mathbb{R}^2 \mid x^2 + y^2 = r\}$. The right side of the last equation is exactly the equation of a circle with radius r about the origin unless $r = 0$. If $r = 0$, there is exactly one point (x, y) such that $x^2 + y^2 = 0$ and that is the origin. //

Notice that, in this case, any two relation classes are either disjoint or equal and the union of all of them is the original set $\mathbb{R}^2$. Is that situation always true for any relation on a set A? It is if $\sim$ is an equivalence relation, as we will see in important Theorem 13. On the other hand, this will not be true for every relation.

Exercise 37. Show that for the "greater than or equal to" relation on $\mathbb{R}$ of Exercise 35, that

A. $[0]$ is the non-negative real numbers.
B. Show that there are $a, b \in \mathbb{R}$ such that $[a] \neq [b]$ but $[a] \cap [b] \neq \emptyset$.

Exercise 37B shows that, for some relations, it may not be true that relation classes are either disjoint or equal.

Definition. If $\sim$ is an equivalence relation on A and $a \in A$ then we call $[a]_\sim$ the EQUIVALENCE CLASS of a with respect to $\sim$. Note that every equivalence class is a relation class. Elements in the same equivalence class are said to be EQUIVALENT (under R). We'll drop the subscript $\sim$ on the equivalence class symbol if it's clear which equivalence relation is being used.

Exercise 38. Using the equivalence modulo 3 relation or "congruent modulo 3", what is
A. $[0]$ B. $[1]$ C. $[2]$
D. $[3]$ E. $[4]$ F. $[5]$
G. $[-1]$ H. $[-2]$ I. $[-3]$
Hint: The answer to each part is a set. In fact, each is a subset of $\mathbb{Z}$. We have written $[0]$, for example, to mean $[0]_3$.

One of the ways for setting up rigorous foundations for clock arithmetic and fractions is contained in the next theorem. This result shows that an equivalence relation on A breaks A into a collection of subsets in which any two elements of any given equivalence class are related and no element is in 2 distinct subsets. This result (Theorem 12) is anticipated by Examples 63A and 65.

Theorem 12. *If A is a set and $\sim$ is an equivalence relation on A, and a and b are elements of A then either $[a] \cap [b] = \emptyset$ or $[a] = [b]$.*

Proof. The logical chain of this proof is to assume that $[a] \cap [b]$ is not the empty set and then conclude that $[a] = [b]$. If $[a] \cap [b] \neq \emptyset$, there is an element c of A with $c \in [a]$ and $c \in [b]$. Thus

$$a \sim c \text{ and } c \sim b. \tag{3.1.4}$$

From transitivity, (3.1.4) shows that

$$a \sim b \tag{3.1.5}$$

and we will use that fact to show that the two sets $[a]$ and $[b]$ are equal.

To demonstrate that two sets are equal we need to show that every element of the first is a member of the second and every element of the second is a member of the first.

We will start with $d \in [a]$, an arbitrary element and show that d is also an element of $[b]$. Since $d \in [a]$, $a \sim d$, but (3.1.5) says that $a \sim b$ so, by transitivity and symmetry, $b \sim d$. Thus d is related to b and $d \in [b]$. A similar argument shows that if $m \in [b]$ then $m \in [a]$ and so the proof is established. ∎

Another way of stating this result is two equivalence classes are either equal or they have no elements in common; that is, the equivalence classes are DISTINCT.

Theorem 13. *If A is a set $\sim$ and is an equivalence relation on A, then A can be written as the union of equivalence classes in which no two distinct classes overlap.*

Proof. Since A is the union of all equivalence classes (because any element is equivalent to itself), the theorem follows immediately from Theorem 12. ∎

Problem Set 3.1:

1. What is:

    (a) The set of points related by the absolute value relation to 0?
    (b) The set of points related by the absolute value relation to 1?

2. (a) The set of points related by the norm relation to $(1, 0)$?
    (b) The set of points related by the norm relation to $(-1, 0)$?
    (c) The set of points related by the norm relation to $(0, 0)$?

3. Show that any two odd integers are congruent modulo 2.
4. For the equivalence relations on $\mathbb{Z}$ given by $\equiv_n$ find the number of distinct equivalence classes if

   A. $n = 3$
   B. $n = 4$
   C. $n = 12$
   D. n arbitrary

5. Using the relation $\equiv_2$ what are other names for $[0]$ and $[1]$?
6. Using the relation $\equiv_{12}$ is there anything in the room you are in which looks like the collection $\{[0], [1], [2], \ldots, [11]\}$? Hint: Is there a clock on the wall?

   Definition. For $a, b \in \mathbb{R}$, we define the LESS THAN RELATION by

   $$a \sim_{\text{less}} b \text{ if } a < b$$

7. Answer the following questions for the less than relation.

   A. What is the relation class of 1?
   B. What is the relation class of 2?
   C. Is Theorem 12 true for $\sim_{\text{less}}$? Why?
   D. Is $\sim_{\text{less}}$ an equivalence relation? Why?

8. Let $\mathscr{L}$ be the set of all lines in $\mathbb{R}^2$ (the plane). We say that for $\ell_1, \ell_2 \in \mathscr{L}$, ℓ_1 is "parallel related" to ℓ_2 if ℓ_1 and ℓ_2 are parallel. Is being parallel a symmetric, reflexive, or transitive relation? (Hint: Look up the definition of parallel lines from a geometry book).
9. Is the "norm relation" from the definition at the beginning of this section an equivalence relation? Why?
10. If the relation R on the set A is reflexive, prove that, for any $a \in A$, $a \in [a]_R$.
11. If $\mathscr{T}$ is the set of all triangles in the plane and R is defined for T_1 and T_2 as $T_1 \sim_s T_2$ if the angle measures of T_1 are equal to the corresponding angle measures of T_2, show that $\sim_s$ is an equivalence relation. What do you think the letter s might stand for in the symbol $\sim_s$?
12. Suppose that $a, b \in \mathbb{Z}$. Define a relation R is by aRb if $a^2 - b^2$ is divisible by 9. Is R an equivalence relation of $\mathbb{Z}$?
13. If Q is the relation of $\mathbb{R}^2$ given by $Q((x_1, y_1), (x_2, y_2))$ is true if $x_1 = x_2$

   (a) Prove that Q is an equivalence relation.
   (b) What are the equivalence classes for Q?

3.2 Modular (Clock) Arithmetic Through Equivalence Relations

Overview In the last section, we gave the general definition of equivalence relations on a set A, showed how relations determine equivalence classes, and how the equivalence classes partition the set A into a collection of subsets which were either identical or pairwise disjoint (Theorem 13). We will now present, in the next two sections, two striking examples of the use of equivalence relations. This section provides a rigorous foundation for modular or clock arithmetic and the next one provides the rigorous basis for fractions.

We will devote this section to the example of the integers $\mathbb{Z}$ with the equivalent modulo n and visualize the equivalence classes as a clock. Working with the integers and the "equivalent modulo n" is called MODULAR ARITHMETIC.

We will use the usual clock (using equivalent modulo 12) as an example during the section. With the experience you have from the preceding section, you will be able to show the following

Example 66. Let $n = 12$. Each equivalence class (under equivalence modulo 12) is either

$$[0], [1], [2], [3], \ldots, [10], [11]$$

where we have dropped the subscript 12.

Solution. From Theorem 12, we need to show that any two equivalence classes are equal or disjoint. So we really need only show that any integer is in one of these classes. We carried out the process for $n = 3$ in Example 63B and the procedure is similar.

Let $a \in \mathbb{Z}$. The Division Algorithm shows that there are $q, r \in \mathbb{Z}$ such that

$$a = 12q + r \text{ where } 0 \leq r < 12.$$

Since $q \in \mathbb{Z}$ and $a - r = 12q$ is divisible by 12, $a \equiv_{12} r$. The proof is complete because the remainder r lies between 0 and 11 (inclusively).   //

This result allows us to picture the collection of equivalence classes (mod 12) as a clock. (We could write [12] instead of [0] as they are equal sets but it is better to use [0] as indicated by the division algorithm, and soon we'll see why.) We are thus led to make the following (Fig. 3.1).

Definition. The set of equivalence classes of $\mathbb{Z}$ with respect to the "equivalence relation modulo n" is called the INTEGERS MODULO n and is written $\mathbb{Z}_n$.

Notation. $\mathbb{Z}_n$ has n elements: equivalence classes of $0, 1, \ldots, n-1$. To simplify the notation, we shall write $\overline{k}$ to represent the equivalence class $[k]$, under $\sim_n$, of the integer k since n is fixed. Of course, $\overline{k} = \overline{k + n}$ since $k \sim_n k + n$. We continue to write $a \equiv b \bmod n$ for $a \sim_n b$.

Fig. 3.1 Usual 12-clock

Fig. 3.2 The n-clock

A clock with n numerals, labeled $\overline{0}, \overline{1}, \ldots, \overline{n-1}$, can be thought of as a physical representation of $\mathbb{Z}_n$. Figure 3.2 has such a clock (and you've seen many with $n = 12$ although 12 (not zero) is used on the clock's face).

The point of these definitions is that we can put clock arithmetic on a rigorous foundation; that is, we can now add and multiply clock numerals. Here is an example on the 12-clock.

When we ask people what time it is 5 h after 11, they will either count it on a clock or their fingers (having 12 fingers would make that easier!) or occasionally say that "$5 + 11 = 16$ but we are working on a 12 h clock so 16 is really 4." If pressed, they may actually say that 4 comes about because $16 - 12 = 4$. What these folks are really doing is modular arithmetic with $n = 12$.

In our language, with $n = 12$, we are asking them to add 5 to $[11]$ (not 11) and they are thinking $[11] + [5] = [16] = [4]$ where the last equality holds because $16 \sim_{12} 4$ or $16 \equiv 4 \bmod 12$.

How would we guess to multiply on a 12 h clock? What would, say, 3 times eleven o'clock be? We regard multiplying as repeated addition so

$$3 \times [11] = [11] + [11] + [11] = [33] = [9]$$

since $33 \equiv 9 \bmod 12$. Thus, we might guess that we should define

$$[3] \times [11] = [33].$$

The outline above motivates the definitions we now write down. The definitions of multiplication and addition give the structure of arithmetic on a clock.

Definition. If $\overline{k}$ and $\overline{\ell} \in \mathbb{Z}_n$, then the PRODUCT of $\overline{k}$ and $\overline{\ell}$ is defined to be

$$\overline{k}\,\overline{\ell} = \overline{k\ell} = [kl]. \tag{3.2.1}$$

The SUM of $\overline{k}$ and $\overline{\ell}$, is defined to be

$$\overline{k} + \overline{\ell} = \overline{k+\ell} = [k+\ell]. \tag{3.2.2}$$

The product and sum above define the MODULAR ARITHMETIC of $\mathbb{Z}_n$.

Is there anything we must do with these definitions before we start to use them? Most would say "no", but there really is! We'll use addition as an example. What is $[2] + [3]$ when $n = 4$. The definition says that $\overline{2} + \overline{3} = \overline{5}$, or

$$[2] + [3] = [2 + 3] = [5]. \tag{3.2.3}$$

But (and this is the key point), there are many ways to represent $\overline{2}$ (or $[2]$). For example, both 2 and 6 represent the same equivalence class $\overline{2}$ since $2 \in \overline{2}$ and $6 \in \overline{2}$. Thus we could have written

$$[6] + [3] = [6 + 3] = [9] \tag{3.2.4}$$

instead of (3.2.3). In order for addition to be WELL DEFINED, it must be that the answers from (3.2.3) and (3.2.4) must be the same, *and they are*. After all, although 9 and 5 are different integers ($5 \neq 9$), *the question is whether the set* $[5]$ *is equal to the set* $[9]$ *or not under the equivalent modulo 4* (not whether $5 = 9$). Since $9 - 5$ is a multiple of 4, $9 \equiv 5$ mod 4 and so they both define the same equivalence class, thus $[9] = [5]$. At least in this case, addition is WELL DEFINED, that is, the operation for addition does not depend on the representation chosen for the equivalence classes. But, we must prove that both addition and multiplication are well defined for all $\overline{k}$ and $\overline{\ell}$. The definition of "well defined" is contained in the statement of Proposition 8. Roughly, it says that addition (or multiplication) will give you the same equivalence class no matter which representatives you use to add (or multiply).

Proposition 8. *For the integers modulo n on* $\mathbb{Z}$*, if* $[k_1] = [k_2]$ *and* $[\ell_1] = [\ell_2]$ *then*

A. $(k_1 + \ell_1) \equiv_n (k_2 + \ell_2)$ *and thus* $[k_1 + \ell_1] = [k_2 + \ell_2]$ *(addition is well defined)*.
B. $(k_1 \ell_1) \equiv_n (k_2 \ell_2)$ *and thus* $[k_1 \ell_1] = [k_2 \ell_2]$ *(multiplication is well defined)*.

Proof. We leave Part B as Problem 2 of Problem Set 3.2. For Part A, since $[k_1] = [k_2]$, $k_1 \equiv_n k_2$ and so there is an integer p with $k_1 - k_2 = pn$. Similarly, there is an integer q with $\ell_1 - \ell_2 = qn$. Thus

$$(k_1 + \ell_1) - (k_2 + \ell_2) = (p + q)n$$

and therefore $k_1 + \ell_1$ and $k_2 + \ell_2$ differ by a multiple of n. We conclude from the definition of integers modulo n that $(k_1 + \ell_2) \equiv_n (k_2 + \ell_2)$ and so those two elements determine the same equivalence class, that is, $[k_1 + \ell_1] = [k_2 + \ell_2]$. ∎

We may now use Eqs. (3.2.1) and (3.2.2) to define addition and multiplication modulo n because Proposition 8 says that the ambiguity of those equations (picking the representative of the equivalence class) does not matter in the end.

What is fascinating is to watch children work with clock arithmetic. They will instinctively realize that $10 + 3 = 13 = 1$ meaning for us that $\overline{10} + \overline{3} = \overline{13} = \overline{1}$ in a 12 h clock and so clock (or modular) arithmetic is intuitive. If you introduce a 5-clock ($n = 5$) or an 8-clock ($n = 8$), they will adapt immediately.

We have learned that systems in which we can add and multiply have other properties, such as associative, commutative, and distributive laws. These properties carry over immediately to modular arithmetic, (although others may not). There is also an analogy to 0 and 1 in modular arithmetic as you will show in Problems 6 and 7 at the end of this section.

Notation. We will use "zero in $\mathbb{Z}_n$" to mean the equivalence class of 0 and "one in $\mathbb{Z}_n$" the equivalence class of 1.

Definition. MODULAR ARITHMETIC, or ARITHMETIC MODULO n, is the set $\mathbb{Z}_n$ together with addition and multiplication as defined in Eqs. (3.2.1) and (3.2.2). The ZERO ELEMENT of $\mathbb{Z}_n$ is $\overline{0}$ and the ONE ELEMENT of $\mathbb{Z}_n$ is $\overline{1}$.

We will now examine whether the usual rules of arithmetic (for the integers $\mathbb{Z}$) are also obeyed in modular arithmetic. The differences between $\mathbb{Z}_n$ and $\mathbb{Z}$ often can make for some exciting classes in middle and elementary school. Of course, for any two integers $a, b \in \mathbb{Z}$ if $ab = 0$ then one (or both) are zero. *That statement need not be true in modular arithmetic.* For example, if $n = 12$ then $(\overline{2})(\overline{6}) = \overline{0}$ but $\overline{2} \neq \overline{0}$ and $\overline{6} \neq \overline{0}$. Thus it is possible for there to be elements $\overline{a}, \overline{b} \in \mathbb{Z}_n$ with $\overline{ab} = \overline{0}$ but $\overline{a} \neq \overline{0}$ and $\overline{b} \neq \overline{0}$. When does this happen?

Proposition 9. *If $n > 1$ is not a prime number then in $\mathbb{Z}_n$ there are non-zero elements $\overline{k}$ and $\overline{\ell}$ with $\overline{k\ell} = \overline{0}$.*

Proof. Since n is not prime, there are integers k and ℓ with $k\ell = n$ and neither k nor ℓ is equal to n. Thus $\overline{k\ell} = [k][\ell] = [k\ell] = [n] = [0] = \overline{0}$ Since k and ℓ are between 0 and n, neither of them is a multiple of n and so $\overline{k} \neq \overline{0}$ and $\overline{\ell} \neq \overline{0}$.　∎

The converse is also true.

Proposition 10. *Suppose that for a given $n > 1$, $\mathbb{Z}_n$ has the property P:*

$$\overline{k} \text{ and } \overline{\ell} \in \mathbb{Z}_n \text{ and } \overline{k\ell} = \overline{0}, \text{ then either } \overline{k} = \overline{0} \text{ or } \overline{\ell} = \overline{0}.$$

If property P holds for all $\overline{k}$ and $\overline{\ell}$, then n is a prime number.

Proof. We will assume n has property P and must prove that n is prime. Suppose that $n = ab$ where $a, b \in \mathbb{Z}$. Usually we have shown that either $a = 1$ or $b = 1$. In this case, we will do what is logically the same, show that either $a = n$ or $b = n$. Since $n = ab$ then $[n] = [a][b]$ but $[n] = \overline{0}$, so

$$\overline{0} = \overline{a}\overline{b}.$$

Property P says that in this case either $\overline{a} = \overline{0}$ or $\overline{b} = \overline{0}$. If $\overline{a} = \overline{0}$ then a is a multiple of n which means that $a = n$ since $ab = n$ and both a and b are integers. The same logic applies to b. Thus n must be a prime number.　∎

What we've seen is the fascinating result that in n-clock arithmetic it is possible that two non-zero "numbers" can multiply to "zero" exactly when n is a composite number (that is, not prime). Try it yourself with $n = 6$.

We will present division in $\mathbb{Z}_n$ in a series of Exercises at the end of this section, separating carefully out two different cases, one in which n is a prime number and the other in which n is a composite number.

We will now give an example of how an equivalence relation can be used to describe a geometric shape (a circle). It gives us an opportunity to see sets and equivalence relations applied to topics (like geometry) other that those in number theory. In fact, the study of rotations and translations can also be phrased in the language of equivalence classes. As an aside, we may define two triangles as being CONGRUENT if there is a rotation or a translation of the plane or a combination of these which carries one triangle to the other. The definition of congruence is the same concept as the idea of congruence of triangles from middle and high schools.

Definition. Let $\mathbb{R}$ be the real numbers and put the relation

$$r_1 \sim_{\mathbb{Z}} r_2 \text{ if } r_1 - r_2 \in \mathbb{Z}$$

on $\mathbb{R}$. This relationship is called EQUALS MOD $\mathbb{Z}$ or EQUALITY MODULO $\mathbb{Z}$ or EQUALITY MOD 1.

The set of equivalence classes mod $\mathbb{Z}$ is usually written $\mathbb{R}/\mathbb{Z}$ and read "$\mathbb{R}$ mod $\mathbb{Z}$". The next two exercises describe all the equivalence classes of $\mathbb{R}/\mathbb{Z}$.

How can we visualize the set of equivalence classes mod 1 (called "equivalence classes mod 1"), $\mathbb{R}/\mathbb{Z}$? It is tempting to think of $\mathbb{R}/\mathbb{Z}$ as the interval between 0 and 1, but that is incorrect because of [0] and [1]. Put another way, what does the set of equivalence classes mod 1 look like? For each real number between 0 and 1 there is a unique equivalence class and every equivalence class is represented except the equivalence class of 0 (see Problems 10 and 11).

Unfortunately, we can't identify the set of equivalences with the numbers between 0 and 1 *inclusive* because 0 and 1 represent the same element of the equivalence classes mod 1 because 0 and 1 give the same equivalence class, which is $\mathbb{Z}$ (Problem 10). Note that if $r > 0$ and $r \to 0$ then $[r]$ approaches $[0] = \mathbb{Z}$ (loosely speaking).

Note that if $s < 1$ and $s \to 1$ then $[s] \to [1] = \mathbb{Z}$. Thus we may visualize the equivalence classes mod 1 as the interval $0 \leq r < 1$ but with the endpoints (0 and 1) identified. Try this construction with a string yourself and you get a circle with circumference 1.

The result of the identification is that $\mathbb{R}/\mathbb{Z}$ "looks like" a circle. This little example shows how the properties of numbers can turn into questions and problems of geometry (or topology to give the above figure a different field). We've just shown a way of constructing the circle by using equivalence classes (of course, there are many other ways). Relations, especially equivalence relations, permeate modern mathematics. As a thought problem, how would you define addition or multiplication on the set of equivalence classes mod $\mathbb{Z}$ (the circle)? Prove that your definition is well defined.

Problem Set 3.2:

1. Do the following clock arithmetic.

   (a) On a 5-clock, what is 3 h after [2]? What is 8 h after [2]?
   (b) On a 12-clock, what is 5 h before 3 (Mathematically, what is $[3] - [5]$)?
   (c) How many times can you add [8] to itself on a 12-clock to get the zero in $\mathbb{Z}_{12}$, or is this impossible?

2. Show that in modular arithmetic $\overline{0}$ is the set of multiples of $\overline{n}$.
3. Prove that multiplication in $\mathbb{Z}_n$ is well defined (that is, prove Proposition 8B).
4. Prove, in modular arithmetic, that $(\overline{a} + \overline{b})\overline{c} = \overline{a}\overline{c} + \overline{b}\overline{c}$.
5. In this exercise, we define subtraction in $\mathbb{Z}_n$ and ask you to prove that it is well-defined. If $\overline{k}$ and $\overline{\ell}$ are elements of $\mathbb{Z}_n$, we define

$$\overline{k} - \overline{\ell} = \overline{k - \ell}.$$

   (a) Suppose that $\overline{k_1} = \overline{k_2}$ and $\overline{\ell_1} = \overline{\ell_2}$. Show that $k_1 - \ell_1$ is equivalent to $k_2 - \ell_2$.
   (b) Conclude from Part A that $\overline{k_1 - \ell_1} = \overline{k_2 - \ell_2}$.

   Compare your proof with that of Proposition 8.

6. In modular arithmetic, show that the only element of $\mathbb{Z}$ that "looks like" the additive identity is $\overline{0}$. More precisely, prove that if

$$\overline{x} + \overline{a} = \overline{a} \text{ for all } \overline{a} \in \mathbb{Z}_n$$

   then $\overline{x} = \overline{0}$.

7. In modular arithmetic, show that the only element of $\mathbb{Z}_n$ that "looks like" the multiplicative identity is $\overline{1}$. More precisely, prove that if

$$\overline{x}\,\overline{a} = \overline{a} \text{ for all } \overline{a} \in \mathbb{Z}_n$$

   then $\overline{x} = \overline{1}$.

8. Show that in $\mathbb{Z}_{12}$

$$\overline{x} + [12] = \overline{x} \text{ for all } \overline{x} \in \mathbb{Z}_{12}.$$

   Does this contradict Problem 3?

9. Prove that equality mod 1 is an equivalence relation on $\mathbb{R}$.

10. Show that $[0] = \mathbb{Z}$ and $[1] = \mathbb{Z}$. Inequality mod $\mathbb{Z}$.
11. If $r \in \mathbb{R}$ then, inequality mod 1,

   (a) $[r] = \mathbb{Z}$ or
   (b) there is an s with $0 < s < 1$ such that $[r] = [s]$.

Problem Set 3.2: Division

The purpose of these exercises is the explore the notion of division in $\mathbb{Z}_n$. There is a significant difference in the results that follow depending on whether n is a prime number or not. We first give two examples ($n = 5$ and $n = 12$) to motivate the definitions and then deal with the general theory.

1 Div. On a 5-clock (5 is prime), show that for every $\overline{x} \in \mathbb{Z}_5$, except $[0]$, we can add x to itself enough times to get $\overline{1}$ (Hint: list all of the elements of $\mathbb{Z}_5$ and try it individually).
2 Div. On a 12-clock (12 is not prime), is it true that for every $\overline{x} \in \mathbb{Z}_{12}$, we can add $\overline{x}$ to itself enough times to get $\overline{1}$? For which $\overline{x}$ can we and for which can we not?

Definition. If $\overline{x} \in \mathbb{Z}_n$ then we say that $\overline{x}$ has a MULTIPLICATIVE INVERSE if there is a $\overline{y} \in \mathbb{Z}_n$ such that $\overline{x}\,\overline{y} = \overline{1}$.

You have shown that every non-zero element of $\mathbb{Z}_5$ has a multiplicative inverse but that this is not true for $\mathbb{Z}_{12}$.

3 Div. Analogous to Problem 1 Div, show that if n is a prime number, every non-zero element $\overline{x}$ has a multiplicative inverse in $\mathbb{Z}_n$.
4 Div. Analogous to Problem 2 Div, show that if n is not a prime number, then there exists a $\overline{x} \in \mathbb{Z}_n$ such that $\overline{x} \neq 0$ and $\overline{x}$ does not have a multiplicative inverse.
5 Div. Let n be a prime number and $\overline{a}, \overline{b} \in \mathbb{Z}_n$ with $\overline{a} \neq \overline{0}$. Prove that there is one and only one $\overline{x} \in \mathbb{Z}_n$ such that $\overline{a}\,\overline{x} = \overline{b}$. (That is the linear equation can be solved.)
6 Div. Let n be a composite number. Show that there exists $\overline{a}, \overline{b} \in \mathbb{Z}_n$ with $\overline{a} \neq \overline{0}$ such that there does not exist a $\overline{x}$ with $\overline{a}\,\overline{x} = \overline{b}$. (Hint: Look at Problem 4 Div.)

3.3   Fractions Through Equivalence Relations

One of the truly subtle concepts of the mathematics of late elementary school and early middle school is that of fractions. In most books for future teachers the definition of a fraction is given as a "part of the whole" and operations on fractions given as rules and motivated by representations of fractions by colored regions. (pizza model), fraction strips, set model, or the number-line model, for example (see Chapter 6, Long et al. [6]). The criteria for a successful representation of a fraction is a clear answer to each of the following questions:

- What is a unit?
- Into how many equal parts has the unit been subdivided?
- How many of these parts are under consideration?

For the fraction $\frac{a}{b}$, b is the answer to the second bullet and a is the answer to the third bullet.

We believe very strongly in the intuitive powers of pictures, connections, and representations. However, there also needs to be a logically consistent basis for any part of mathematics and the concept of fractions should be no exception. A good overview of how mathematicians present fractions to future elementary school teachers is contained in McCrory (2006) (See [7]). Ultimately each approach is circular or assumes some geometric properties of the line. We are not arguing that fractions should be taught to either future elementary teachers or to the children using equivalence relations as presented in this section. Rather, this section is intended for those who need to understand deeply the logical foundations on which the arithmetic of fractions lies. Those people include math specialists in elementary school, future middle or high school teachers, mathematically sophisticated high school students, or math majors.

We now define a set F and an equivalence relation on F. The equivalence classes of F (with regard to this relation) will give a precise definition of what a fraction is. We then proceed to show that the usual definitions of addition and multiplication of fractions are well-defined in this rigorous treatment. Our approach will follow the logical argument of the previous section for adding and multiplying in modular arithmetic. We'll see immediately (in Eq. (3.3.1)) that the "equivalent as fractions" relation mimics the usual statement that the fractions $\frac{a_1}{b_1} = \frac{a_2}{b_2}$ if and only if $a_1 b_2 = a_2 b_1$. We'll also see that, just as in modular arithmetic, we will need to show that each operation on fractions is WELL DEFINED.

Definition. Let F be the set

$$F = \{(a, b) \mid a \text{ and } b \text{ are integers and } b \neq 0\}$$

If $\alpha = (a_1, b_1)$ and $\beta = (a_2, b_2)$ are elements of F then we say that α and β are EQUIVALENT AS FRACTIONS if

$$a_1 b_2 = a_2 b_1 \tag{3.3.1}$$

and write $\alpha \sim_F \beta$

Proposition 11. *The relation $\sim_F$ is an equivalence relation on the set F.*

Proof. It is easy to see that $\sim_F$ is both reflexive and symmetric. To show that $\sim_F$ is transitive assume that $\alpha = (a_1, b_1)$, $\beta = (a_2, b_2)$ and $\gamma = (a_3, b_3)$ are elements of F with $\alpha \sim_F \beta$ and $\beta \sim_F \gamma$. We have

$$a_1 b_2 = a_2 b_1 \text{ and } a_2 b_3 = a_3 b_2.$$

We must show that $\alpha \sim_F \gamma$ which is the same as $a_3 b_1 = a_1 b_3$. Since $b_2 \neq 0$, we may use the previous equalities to obtain

$$a_3 b_1 = \frac{a_2 b_3}{b_2} b_1 = \frac{(a_2 b_1) b_3}{b_2} = \frac{(a_1 b_2) b_3}{b_2} = a_1 b_3.$$

Thus $\alpha \sim_F \gamma$ by the definition of equivalent as fractions. ∎

Definition. The SET OF FRACTIONS $\mathbb{Q}$ is the set of equivalence classes of F with respect to the equivalence relation $\sim_F$. It is written as

$$\mathbb{Q} = F / \sim_F$$

It is important to remember that an element of $\mathbb{Q}$ is actually a set of pairs of integers (where the second integer isn't zero). A FRACTION is an element of $\mathbb{Q}$. This means that a fraction, in our approach, is a set of pairs of integers in which the second coordinate is never zero.

Of course, this is a very formal definition of fraction, depending only on the construction of the set of integers, $\mathbb{Z}$, and some elementary set theory. How does this construction relate the definition of fractions from elementary school?

Notation. Throughout the remainder of this section, $[\alpha]$ will mean the equivalence class of α with respect to $\sim_F$.

If $\alpha = (a, b) \in F$ then the fraction $\frac{a}{b}$ is the equivalence class of (a, b). Thus $\frac{a}{b} = [(a, b)] \in \mathbb{Q}$ (note that b cannot be zero because of the way the set F is defined).

The next example may be difficult at first, but understanding it thoroughly will make the rest of the material in this section easier.

Example 67. Let $\alpha = (1, 2) \in F$. What does $[\alpha]$ look like?

Solution. If $\beta \in [\alpha]$ and $\beta = (a, b)$ then $(1, 2) \sim_F \beta$ so that (3.3.1) is

$$1 \cdot b = a \cdot 2 \text{ or } b = 2a.$$

In other words, there are infinitely many pairs in $[\alpha]$ such as

$$(-2, -4), (-1, -2), (1, 2), (2, 4), (3, 6), (4, 8), \ldots \in [(1, 2)]. \qquad (3.3.2)$$

In the language of children, each of these are the fraction written as $\frac{1}{2}$. //

Notation. If $(a, b) \in F$ then the fraction $\frac{a}{b}$ is the fraction equivalence class of (a, b). Thus $\frac{a}{b} = [(a, b)] \in \mathbb{Q}$. (Note that b cannot be zero because of the way in which the set F is defined.)

Example 68. Show that $\frac{-1}{-2} = \frac{1}{2} = \frac{2}{4} = \frac{3}{6} = \frac{4}{8} = \cdots$

Solution. $\frac{1}{2}$ is the *set* $[(1,2)]$ which includes (see (3.3.2)) the elements $(-1,-2)$, $(2,4)$, $(3,6)$, $(4,8)$, etc. Because of Theorem 12, since $(2,4) \in [(1,2)]$, it must be that $[(2,4)] = [(1,2)]$. Each of the other fraction equalities are proven in the same way.   //

Good news: our formalism actually shows that a fraction can be represented by many different pairs of numbers, just as in grade school!

Before we prove the next result, let's look at what is involved in showing that two fractions are equal. Since a fraction is a set, we should prove that each of the pairs are in the same set. Fortunately, thanks to Theorem 12, we only need to prove that there is one element in one of the sets which is also in the other *and that is enough.* Now would be a good time to review the consequences of Theorem 12.

Here's another result which is very familiar to you and a key to the conceptual understanding of fractions.

Proposition 12. *If n is a non-zero integer and $\frac{a}{b}$ is a fraction then*

$$\frac{a}{b} = \frac{na}{nb} \qquad\qquad (3.3.3)$$

Proof. Given $\frac{a}{b} \in \mathbb{Q}$, we can write it as $\frac{a}{b} = [(a,b)]$. (Of course, we could have chosen many other representatives of the equivalence class $\frac{a}{b}$.) Let

$$\beta = (na, nb) \in F.$$

Proving this proposition means answer the question, is β fraction equivalent to (a,b)? This question can be restated as asking if

$$(a,b) \sim_F (na, nb) = \beta.$$

From the definition of $\sim_F$ (see (3.3.1)) this happens exactly when

$$a(nb) = (na)b,$$

which is always true. Since (a,b) is fraction equivalent to (na, nb), they determine the same equivalence class and hence the same fraction. Therefore $\frac{a}{b} = \frac{na}{nb}$. ∎

We now turn our attention to operations with fractions, specifically addition and multiplication. As in the case of modular arithmetic, we must show that the definition we will use for addition and multiplication does not depend on the choice of the representative of the equivalence class. The usual way of defining addition and multiplication of fractions is

$$\left(\frac{a_1}{b_1}\right)\left(\frac{a_2}{b_2}\right) = \frac{a_1 a_2}{b_1 b_2} \qquad\qquad (3.3.4)$$

$$\frac{a_1}{b_1} + \frac{a_2}{b_2} = \frac{a_1 b_2 + a_2 b_1}{b_1 b_2} \qquad\qquad (3.3.5)$$

where a_1, b_1, a_2, b_2 are integers with $b_1 \neq 0$ and $b_2 \neq 0$. What is never explained in elementary books about arithmetic is why these definitions actually make sense. After all, $\frac{1}{2}$ and $\frac{2}{4}$ are the same quantity. What would happen if I chose $\frac{1}{2}$ for the first fraction and you chose $\frac{2}{4}$, and I picked $\frac{2}{3}$ and you $\frac{6}{9}$ for the second? Of course, the answers $\left(\frac{1}{2}\right)\left(\frac{2}{3}\right)$ and $\left(\frac{2}{4}\right)\left(\frac{6}{9}\right)$ are the same but there is actually something to prove.

The issue is whether the operation of multiplication is WELL DEFINED. When doing modular arithmetic, we also needed to prove that addition and multiplication were well defined. We first ask you to do two exercises with specific fractions as preparation for the general result (the first of which is also left as an exercise).

Exercise 39. Remember $\frac{2}{3}$ is a set and so is $\frac{1}{5}$. Consider $\frac{2}{3} = [(2, 3)]$ and $\frac{1}{5} = [(1, 5)]$. Pick any two elements $\alpha_1 = (a_1, b_1)$ and $\alpha_2 = (a_2, b_2)$ of $\frac{2}{3}$ and any two elements of $\beta_1 = (c_1, d_1)$ and $\beta_2 = (c_2, d_2)$ of $\frac{1}{5}$ and show that

A. $(a_1c_1, b_1d_1) \sim_F (a_2c_2, b_2d_2)$
B. $[(a_1c_1, b_1d_1)] = \frac{2}{15}$.

In the language of elementary school, we've asked you to show that

$$\left(\frac{2}{3}\right)\left(\frac{1}{5}\right) = \frac{2}{15}$$

independent of what representatives of the equivalence classes are chosen.

Theorem 14. *Suppose (a_1, b_1), (a_2, b_2), (c_1, d_1), (c_2, d_2) are elements of F. If*

$$(a_1, b_1) \sim_F (a_2, b_2) \text{ and } (c_1, d_1) \sim_F (c_2, d_2) \tag{3.3.6}$$

then

$$(a_1c_1, b_1d_1) \sim_F (a_2c_2, b_2d_2) \tag{3.3.7}$$

and

$$\frac{a_1d_1 + c_1b_1}{b_1d_1} \sim_F \frac{a_2d_2 + b_2c_2}{b_2d_2}. \tag{3.3.8}$$

Proof. We'll leave the first result (3.3.7) as Exercise 40. We'll now prove (3.3.8). We know that, because the pairs are fraction equivalent,

$$a_1b_2 = a_2b_1 \text{ and } c_1d_2 = c_2d_1. \tag{3.3.9}$$

Thus the left hand side of (3.3.8) is

$$\frac{a_1d_1 + c_1b_1}{b_1d_1} = \frac{b_2d_2(a_1d_1 + c_1b_1)}{b_1b_2d_1d_2} = \frac{(a_1b_2)(d_1d_2) + (c_1d_2)b_1b_2}{b_1b_2d_1d_2}.$$

Using (3.3.9) we find

$$\frac{a_1 d_1 + c_1 b_1}{b_1 d_1} = \frac{(a_2 b_1) d_1 d_2 + (c_2 d_1)(b_1 b_2)}{b_1 b_2 d_1 d_2} = \frac{b_1 d_1 (a_2 d_2 + c_2 b_2)}{b_1 b_2 d_1 d_2}$$

which is exactly the same as the right hand side of (3.3.8).   ∎

Exercise 40. Under the assumptions of Theorem 14, prove Eq. (3.3.7).

Theorem 14 now allows us to make the definition of multiplication and addition of fractions in a formally correct, unambiguous, and rigorous manner since the Theorem says that any two elements of F which are fraction related will give the same fraction equivalence class after addition and multiplication.

Definition. If $\frac{a}{b}$ and $\frac{c}{d}$ are two fractions then the PRODUCT and the SUM of the those fractions are given by

$$\left(\frac{a}{b}\right)\left(\frac{c}{d}\right) = \frac{ac}{bd}$$

and

$$\frac{a}{b} + \frac{c}{d} = \frac{ad + bc}{bd}.$$

Problem Set 3.3:

1. Remember that $\frac{3}{4}$ is a set and so is $\frac{3}{5}$. Consider $\frac{3}{4} = [(3, 4)]$ and $\frac{3}{5} = [(3, 5)]$. Pick two elements $\alpha_1 = (a_1, b_1)$ and $\alpha_2 = (a_2, b_2)$ of $\frac{3}{4}$ and any two elements $\beta_1 = (c_1, d_1)$ and $\beta_2 = (c_2, d_2)$ of $\frac{3}{5}$ and show that:

   (a) $(a_1 c_1, b_1 d_1) \sim_F (a_2 c_2, b_2 d_2)$
   (b) $[(a_1 c_1, b_1 d_1)] = \frac{9}{20}$.

   In the language of elementary school, you have just shown that $(\frac{3}{4})(\frac{3}{5}) = \frac{9}{20}$.

2. Remember that $\frac{1}{4}$ is a set and so is $\frac{3}{4}$. Consider $\frac{1}{4} = [(1, 4)]$ and $\frac{3}{4} = [(3, 4)]$. Pick two elements $\alpha_1 = (a_1, b_1)$ and $\alpha_2 = (a_2, b_2)$ of $\frac{1}{4}$ and any two elements $\beta_1 = (c_1, d_1)$ and $\beta_2 = (c_2, d_2)$ of $\frac{3}{4}$ and show that:

   (a) $(a_1 c_1, b_1 d_1) \sim_F (a_2 c_2, b_2 d_2)$
   (b) $[(a_1 c_1, b_1 d_1)] = \frac{3}{16}$.

3. If $b \neq 0$ and $c \neq 0$ and both are integers, show that

   $$[(0, b)] = [(0, c)].$$

4. If $b \in \mathbb{Z}$ is not zero, we will write

   $$\mathbf{0} = [(0, b)] = \frac{0}{b}.$$

Prove that for any fraction $\frac{c}{d}$,

(a) $0 + \frac{c}{d} = \frac{c}{d}$

(b) $(0)\left(\frac{c}{d}\right) = 0$

Note that the equation of Part (a) shows that 0 is an additive identity. It is easy to see that multiplication and addition are commutative and associative.

5. If a is non-zero, prove that

$$\left[\frac{a}{a}\right] = 1.$$

6. Let $a \neq 0$ and write

$$1 = \left[\frac{a}{a}\right].$$

The fraction 1 is well-defined by Problem 5. Prove that

$$1\left(\frac{c}{d}\right) = \frac{c}{d}$$

for any fraction $\frac{c}{d}$. (This equation shows that 1 is the MULTIPLICATIVE IDENTITY.)

7. Let $\frac{a}{b}$ and $\frac{c}{d}$ be fractions. Show that

(a) $\frac{a}{b} + \frac{c}{d} = \frac{c}{d} + \frac{a}{b}$ (addition is commutative).

(b) $\left(\frac{a}{b}\right)\left(\frac{c}{d}\right) = \left(\frac{c}{d}\right)\left(\frac{a}{b}\right)$ (multiplication is commutative).

8. Describe what it means for addition of fractions to be associative and prove your statement.

9. Describe what it means for multiplication of fractions to be associative and prove your statement.

10. Let $\frac{a}{b}$ be a fraction. Prove that there is a fraction x with

$$\frac{a}{b} + x = 0$$

if and only if $x = \frac{-a}{b}$. This is the statement of existence and uniqueness of ADDITIVE INVERSES OF FRACTIONS.

11. Let $\frac{a}{b}$ be a fraction with $a \neq 0$. Prove that there is a fraction x with

$$\left(\frac{a}{b}\right)x = 1$$

if and only if $x = \frac{b}{a}$. This is the statement of existence and uniqueness of MULTIPLICATIVE INVERSES OF FRACTIONS.

12. Show that multiplication is distributive across addition for fractions. More formally, if a, b, c, d, e and f are integers with $b \neq 0$, $d \neq 0$ and $f \neq 0$ then show that

$$\frac{a}{b} \left(\frac{c}{d} + \frac{e}{f} \right) = \left(\frac{a}{b} \right) \left(\frac{c}{d} \right) + \left(\frac{a}{b} \right) \left(\frac{e}{f} \right).$$

3.4 Integers Modular n and Applications

Overview In Sect. 3.1 the notion of the congruence mod n is introduced. Congruence modulo n is a relation on the set of integers with a related to b if $a \equiv b$ (mod n). It is proved that congruence modulo n is an equivalence relation on the set of integers in Sect. 3.2. This section studies some properties and applications with congruence such as RSA cryptosystem, Universal Product Code (UPC), and International Standard Book Number (ISBN).

Proposition 13. *Assume that $a \equiv b$ (mod n) and $c \equiv d$ (mod n). Then*

A. $a + c \equiv b + d$ *(mod n)*
B. $a - c \equiv b - d$ *(mod n)*
C. $ac \equiv bd$ *(mod n)*

Proof. By assumption $n|(a - b)$ and $n|(c - d)$. Thus $n|(a - b) + (c - d)$ which implies that $n|(a + c) - (b + d)$ which means that $a + c \equiv b + d$ (mod n). This proves A. Similarly it is easy to show B. To prove C, we note that $n|(a - b)$ implies that $n|c(a - b)$, so that $n|(ac - bc)$, i.e., $ac \equiv bc$ (mod n). Similarly, since $n|(c - d)$, we have $n|b(c - d)$ which implies $n|(bc - bd)$, i.e., $bc \equiv bd$ (mod n). Therefore $ac \equiv bd$ (mod n). ∎

Remark. Congruence acts like equality in many ways.

An elementary induction argument allows us to have

Corollary 5. *Assume that $a_1 \equiv b_1$ (mod n), $a_2 \equiv b_2$ (mod n), $\cdots$, $a_m \equiv b_m$ (mod n). Then*

$$a_1 + a_2 + \cdots + a_m \equiv b_1 + b_2 + \cdots + b_m \ (mod \ n)$$

and

$$a_1 a_2 \cdots a_m \equiv b_1 b_2 \cdots b_m \ (mod \ n).$$

Corollary 6. *If $a \equiv b$ (mod n), then $a^k \equiv b^k$ (mod n) for any positive integer k.*

Note: If $a \equiv b$ (mod n), then $ac \equiv bc$ (mod n) for any integer c by Proposition 13C. However, the converse is not true. As an example, $4 \cdot 3 \equiv 2 \cdot 3$ (mod 6), but $4 \not\equiv 2$ (mod 6).

Proposition 14. *If $ac \equiv bc$ (mod n) and* GCD$(c, n) = 1$, *then $a \equiv b$ (mod n).*

Proof. By hypothesis, we have $n|(ac - bc)$ and so $n|(a - b)c$. But GCD$(c, n) = 1$, thus $n|(a - b)$, i.e., $a \equiv b$ (mod n). ∎

Example 69. $5 \cdot 3 \equiv 1 \cdot 3$ (mod 4) and GCD$(3, 4) = 1$ implies that $5 \equiv 1$ (mod 4).

Example 70. What is the remainder when 2^{20} is divided by 41?

Solution. Since $2^5 \equiv -9$ (mod 41), by Proposition 13 C we have

$$2^{10} = 2^5 \cdot 2^5 \equiv (-9)(-9) \equiv 81 \equiv -1 \text{ (mod 41)},$$

and

$$2^{10} \cdot 2^{10} \equiv (-1)(-1) \equiv 1 \text{ (mod 41)},$$

so that $2^{20} \equiv 1$ (mod 41). Therefore the remainder is 1 when 2^{20} is divided by 41.//

Example 71. What is the remainder when 1997^{10} is divided by 19?

Solution. Since $1997 \equiv 2$ (mod 19), by Proposition 13C we have

$$1997^{10} \equiv 2^{10} \text{ (mod 19)}.$$

Now,

$$2^4 \equiv 16 \equiv -3 \text{ (mod 19) and } 2^8 = 2^4 \cdot 2^4 \equiv (-3)(-3) \equiv 9 \text{ (mod 19)}.$$

Thus

$$2^8 \cdot 2^2 \equiv 9 \cdot 4 \equiv 36 \equiv 17 \text{ (mod 19)}.$$

Therefore, the remainder is 17. //

3.4.1   RSA Cryptosystem

Cryptography is the study of sending and receiving secret messages. The goal of cryptography is to send messages across a channel (computers, electronic media) so that only the intended receiver of the message can read it. This is done by enciphering the message in such a way that an unauthorized person will find it very difficult to infer the message.

   The message to be sent is called the plaintext. The message after transformation to a secret form is called the ciphertext. The plaintext and the ciphertext are both written in an alphabet, consisting of letters or characters (A,..., Z), digits, punctuation markers, and blanks. A cryptosystem has two parts: <u>encryption</u>, the process of transforming a plaintext to a ciphertext, and <u>decryption</u>, the process of transforming from ciphertext back to a plaintext. The entire process can be summarized in the diagram:

$$\underset{\text{(plaintext)}}{m} \rightarrow \text{encrypting} \rightarrow \underset{\text{(ciphertext)}}{c} \rightarrow \text{decrypting} \rightarrow \underset{\text{(plaintext)}}{m} \qquad (3.4.1)$$

The most widely known public-key cryptosystem is the **RSA cryptosystem** proposed by Rivest et al. [9]. We will need to define the Euler ϕ-function.

Definition. The EULER PHI FUNCTION $\phi(n)$, is the number of positive integers less than n that are relatively prime (or coprime) to n.

Example 72. What is the Euler phi function $\phi(12)$?

Solution. $\phi(12) = 4$, since 1, 5, 7, and 11 are the only integers that are positive, less than 12, and coprime to 12.   //

Algorithm: Key generation for RSA public-key cryptosystem (see Mollin [8]).
Summary: Each entity creates an RSA public key and a corresponding private key.
   Each entity should do the following:

1. Generate two large random (and distinct) primes p and q, each roughly the same size (about 100 digits).
2. Compute $n = pq$ and $\phi(n) = (p-1)(q-1)$. ϕ depends on the two distinct primes p and q.
3. Select a random integer e, $1 < e < \phi(n)$, such that $\text{GCD}(e, \phi(n)) = 1$.
4. Use the Euclidean algorithm to compute the unique integer d, $1 < d < \phi(n)$, such that $ed \equiv 1 \pmod{\phi(n)}$.
5. A's public key is (n, e); A's private key is (d, p, q).

(I) Encryption. B encrypts a message m for A, which A decrypts.
   B should do the following:

1. Obtain A's public key (n, e).
2. Represent the message as an integer m in the range $\{0, 1, \ldots, n\text{-}1\}$.
3. Compute $c \equiv m^e \pmod{n}$.
4. Send c to A.

(II) Decryption. A should do the following:

1. Compute $m \equiv c^d \pmod{n}, 0 \leq m < n$.

Example 73. Let $p = 3, q = 17$. Then $n = pq = 51$ and $\phi(n) = 2 \cdot 16 = 32$. Since $\text{GCD}(e = 11, \phi(n) = 32) = 1$, we may choose $e - 11$ and hence $d = 3$. To encrypt message "G"=07, compute $c \equiv 7^{11} \equiv (7^8)(7^2)(7) \equiv 31 \pmod{51}$. So Bob sends $c = 31$ to Alice. To decrypt the message using $d = 3$, one has $m \equiv 31^3 \equiv 7 \pmod{51}$. Hence $m = 07 = $ "G".

Example 74. Let $p = 29, q = 53$. Then $n = pq = 29 \cdot 53 = 1537$ and $\phi(n) = 28 \cdot 52 = 1456$. Since GCD$(47, 1456) = 1$, we may choose $e = 47$ and hence $d = 31$. $d = 31$ can be computed by the Euclidean algorithm as follows. Since GCD$(47, 1456) = 1$, we calculate

$$1456 = 30 \cdot 47 + 46$$
$$47 = 1 \cdot 46 + 1$$

Thus,

$$1 = 47 - 1 \cdot 46$$
$$= 47 - 1 \cdot (1456 - 30 \cdot 47)$$
$$= 31 \cdot 47 - 1456.$$

To encrypt message "HI"=0809, $c \equiv 809^{47} \equiv 1439 \pmod{1537}$. So Bob sends $c = 1439$ to Alice. To decrypt the message using $d = 31$, Alice has

$$m \equiv 1439^{31} \equiv (-98)^{31} \pmod{1537}$$
$$\equiv (-1)^{31}(98^{31}) \pmod{1537}$$
$$\equiv (-1)^{31}(728) \pmod{1537}$$
$$\equiv -728 \equiv 809 \pmod{1537}$$

Hence the message $m = 809=$ "HI".

Example 75. Let $p = 43, q = 47$. Then $n = pq = 43 \cdot 47 = 2021$ and $\phi(n) = 42 \cdot 46 = 1932$. Since GCD$(e = 61, \phi(n) = 1932) = 1$, we may choose $e = 61$ and hence $d = 1837$. To encrypt message "BE"=0205, $c \equiv 205^{61} \equiv 1422 \pmod{2021}$. So Bob sends $c = 1422$ to Alice. To decrypt the message using $d = 1837$, Alice has

$$m \equiv 1422^{1837} \equiv (-115)^{1837} \equiv 205 \pmod{2021}$$

Hence the message $m = 205=$ "BE".

Exercise 41. Suppose Alice's public key is $(n, e) = (91, 11)$. If you want to send a message containing the single letter P (16), what ciphertext does she receive, and how does Alice decipher it?

3.4.2   *UPC and ISBN (See Gallin and Winters [3], Rosen [10])*

(I). Universal product code (UPC) is now found on most product in grocery
and retail stores. The UPC has 12 digits. The first six digits identify the
manufacturer, the next five identify the product, and the last is a check digit
(see figure).

$$0 \quad 36000 \; 29145 \quad 2$$

If $d_1 d_2 \cdots d_{12}$ is a valid UPC code, then

$$3d_1 + d_2 + 3d_3 + d_4 + \cdots + 3d_{11} + d_{12} \equiv 0 \ (\text{mod } 10). \qquad (3.4.2)$$

Example 76. Is $0 - 21000 - 65897 - 8$ a valid UPC?

Solution. To verify it, we calculate

$$3\cdot 0 + 1\cdot 2 + 3\cdot 1 + 1\cdot 0 + 3\cdot 0 + 1\cdot 0 + 3\cdot 6 + 1\cdot 5 + 3\cdot 8 + 1\cdot 9 + 3\cdot 7 + 1\cdot 8$$

$$\equiv 2 + 3 + 18 + 5 + 24 + 9 + 21 + 8 \equiv 5 + 8 + 5 + 4 + 9 + 1 + 8$$

$$\equiv 3 + 9 + 0 + 8 \equiv 0 \ (\text{mod } 10)$$

Thus, it is a valid UPC.   //

Example 77. Find the missing digit of UPC $021000x58798$.

Solution. We calculate

$$3\cdot 0 + 1\cdot 2 + 3\cdot 1 + 1\cdot 0 + 3\cdot 0 + 1\cdot 0 + 3\cdot x + 1\cdot 5 + 3\cdot 8 + 1\cdot 7 + 3\cdot 9 + 1\cdot 8 \equiv 3x + 6 \equiv 0 \ (\text{mod } 10).$$

Therefore, $3x + 6 \equiv 0 \ (\text{mod } 10)$. Thus $x = 8$.   //

Exercise 42. Find the correct check digit of the UPC code with 07033020118 as its
first 11 digits.

(II). ISBN-10

Since 1972, a book published anywhere in the world has carried a ten-digit code
called an International Standard Book Number (ISBN). An ISBN consists of four
parts: a group code, a publisher code, an identifying number assigned by the

publisher, and a check digit. For example, in the ISBN 0-201-87073-8, the group code (0) denotes that the book was published in an English speaking country (Australia, Canada, New Zealand, South Africa, the United Kingdom, the USA). The next group (201) denotes the publisher, and the third group (87073) denotes this particular book among all those published by that publisher. The final digit (8) is the check digit, which is used to detect errors in copying or transmitting the ISBN.

The check digit has 11 possible values: 0, 1, 2, 3, 4, 5, 6, 7, 8, 9, or X (X represents the number 10). If $d_1 d_2 \cdots d_{10}$ is a valid ISBN, then

$$10d_1 + 9d_2 + 8d_3 + 7d_4 + 6d_5 + 5d_6 + 4d_7 + 3d_8 + 2d_9 + d_{10} \equiv 0 \pmod{11}.$$

Example 78. Is $0 - 534 - 91500 - 0$ a valid ISBN?

Solution. We calculate

$$10 \cdot 0 + 9 \cdot 5 + 8 \cdot 3 + 7 \cdot 4 + 6 \cdot 9 + 5 \cdot 1 + 4 \cdot 5 + 3 \cdot 0 + 2 \cdot 0 + 1 \cdot 0$$
$$= 45 + 24 + 28 + 54 + 5 + 20$$
$$\equiv 1 + 2 + 6 + 10 + 5 + 9$$
$$\equiv 33 \equiv 0 \pmod{11}.$$

Hence, it is a valid ISBN.   //

Example 79. Find the missing digit of ISBN $0 - 67x - 98039 - 1$.

Solution. Again we calculate

$$10 \cdot 0 + 9 \cdot 6 + 8 \cdot 7 + 7 \cdot x + 6 \cdot 9 + 5 \cdot 8 + 4 \cdot 0 + 3 \cdot 3 + 2 \cdot 9 + 1 \cdot 1$$
$$= 54 + 56 + 7x + 54 + 40 + 9 + 18 + 1$$
$$\equiv 10 + 1 + 7x + 10 + 7 + 9 + 7 + 1$$
$$\equiv 11 + 7x + 11 + 23$$
$$\equiv 7x + 1 \equiv 0 \pmod{11}$$

Hence $7x + 1 \equiv 0 \pmod{11}$ implies $7x \equiv -1 \equiv 10 \pmod{11}$. Hence $x = 3$ (since $7 \cdot 3 \equiv 21 \equiv 10 \pmod{11}$). Thus the missing digit is 3.   //

Exercise 43. Find the correct check digit of the ISBN number with 0-673-48582 as its first nine digits.

Problem Set 3.4:

1. What is (a) $6^{100} \pmod 7$ (b) $83^{25} \pmod 7$ (c) $25^{38} \pmod 7$?
2. What is (a) $935^{40} \pmod{13}$ (b) $40^{255} \pmod{13}$ (c) $1997^{10} \pmod{19}$?

3. Let $n \in \mathbb{N}$. Show that $n \equiv 0 \pmod 3$ if and only if the digit sum of n is divisible by 3.
4. Determine whether each of the following pairs is congruent modulo 9.
   (a) 1, 19   (b) $-1, 8$ (c) $-9, 52$.
5. For which positive integers n is each of the following true?

   (a) $28 \equiv 4 \pmod n$
   (b) $152 \equiv 142 \pmod n$

6. Show that $a^2 \equiv 0, 1 \pmod 4$ for all $a \in \mathbb{Z}$.
7. Show that if n is an odd positive integer, then $1 + 2 + 3 + \cdots + (n - 1) \equiv 0 \pmod n$.
8. Find the least positive residues mod 13 of each of the following integers.

   (a) 2^{30}
   (b) 2^{53}

9. Determine whether each of the following twelve digit strings can be the UPC of a product

   (a) 044000045548
   (b) 031604028465

10. Find the missing digit of UPC 021000?58798
11. Find the missing digit of ISBN-10 $0 - 54-? - 16509 - 9$
12. We are using the RSA system. Suppose I chose $(n,e) = (77, 5), (77, 7)$. If you want to send a message "Z" to me, what ciphertext do you actually send? Explain which (n,e) you pick and why.
13. We are using the RSA system. Suppose my $(n,e) = (33, 13)$ and I receive from you a ciphertext message 22. How do I decipher it?
14. We are using the RSA system. Let p= 97, q= 103. Suppose e= 13, 19. Then what is the cipherkey, respectively?

Part II
The Algebra of Polynomials
and Linear Systems

Chapter 4
Polynomials and the Division Algorithm

Introduction At this point, you are quite adept with algebra and, especially, polynomials. However, in this chapter, we will examine polynomials with four goals in mind that are very different from your previous study:

(a) to give the subject a formal consistency (for example, careful definitions of ideas and operations);
(b) to view a polynomial in two different ways such as an indeterminate form or special kind of function in one variable;
(c) to develop the analogy between the set of numbers and the set of all polynomials.

4.1 Addition and Multiplication of Polynomials

Overview In this section we carefully define what a polynomial is and formally define equality, addition and multiplication of polynomials. These operations allow us to create new polynomials from previously known examples in a similar manner to show how we create new integers or real numbers using addition and multiplication of previously known integers or real numbers. The similarities between the set of polynomials and the set of integers will direct us in our study. For the sake of completeness, we include the following definition.

A student's notion of what a function is changes throughout schooling. At first, it's just an assignment, then it becomes a "general rule" f for assigning to each number in its domain another number, and ultimately $f(x)$ is a way to give to each point x in a set (the domain of f) a value, $f(x)$, in another set (the range of f). The formal symbol x can be thought of as an arbitrary or typical value of the variable, or x can be thought of as a quantity which is not determined (an *indeterminate*).

© Springer International Publishing Switzerland 2015
R.S. Millman et al., *Problems and Proofs in Numbers and Algebra*,
DOI 10.1007/978-3-319-14427-6_4

Definition. Let n be a non-negative integer and $a_0, \ldots, a_n$ be real numbers such that $a_n \neq 0$. An expression of the form

$$p(x) = a_n x^n + a_{n-1} x^{n-1} + \cdots + a_2 x^2 + a_1 x + a_0$$

is called a POLYNOMIAL in the indeterminate x with real numbers as COEFFI-CIENTS such as a_0, a_1, $\ldots,a_n$. The coefficient a_0 is called the CONSTANT TERM and a_n is called the LEADING COEFFICIENT of $p(x)$. If $a_n = 1$, then $p(x)$ is called a MONIC polynomial.

For the sake of completeness, we add the special case where all of the coefficients are equal to zero and we say $p(x)$ is the ZERO POLYNOMIAL. A polynomial in which all coefficients are zero except possibly for the constant term is called a CONSTANT POLYNOMIAL. Occasionally, we may emphasize that $p(x)$ is a polynomial over the real numbers $\mathbb{R}$.

Definition. If n is the largest non-negative integer such that $a_n \neq 0$, then we say that the polynomial $p(x)$ has DEGREE n, and write $\deg(p(x)) = n$. The zero polynomial has no degree.

Definition. Two polynomials are EQUAL when they have the same degree and their corresponding coefficients are equal. More precisely, if

$$p_1(x) = a_n x^n + a_{n-1} x^{n-1} + \cdots + a_1 x + a_0$$

and

$$p_2(x) = b_n x^n + b_{n-1} x^{n-1} + \cdots + b_1 x + b_0,$$

then $p_1(x) = p_2(x)$ if and only if $a_0 = b_0, a_1 = b_1, \ldots$, and $a_n = b_n$.

Example 80. If $(3a - b + 5)x^2 + (3a + 2b - 11)x + (c - 3)$ is the zero polynomial, find the values of $a, b,$ and c.

Solution. By the definition of the zero polynomial, each coefficient must be zero. Thus

$$3a - b + 5 = 0, \quad 3a + 2b - 11 = 0, \text{ and } c - 3 = 0.$$

Solving these three equations for $a, b,$ and c gives:

$$a = \frac{1}{9}, \ b = \frac{16}{3}, \text{ and } c = 3.$$

$//$

Example 81. Let $p(x) = ax^4 + bx^3 + cx^2 + dx + e$ be a polynomial with $a, b, c, d, e \in \mathbb{Z}$ and assume

$$3|a| + 4|b - 2| + |c - 1| = 1. \tag{4.1.1}$$

What are the possibilities for $p(x)$?

Solution. Since the right hand side of Eq. (4.1.1) is 1 and each of a, b, c, d, e are integers, there is a limited number of possibilities and we proceed by inspection. In particular, $3|a| > 1$ implies that $a = 0$. Similarly, $|b - 2| = 0$ and $|c - 1| = 1$. Thus Eq. (4.1.1) implies that:

$$a = 0, \ b = 2, \text{ and } c = 0 \text{ or } 2.$$

Since there are no conditions on d or e, the possibilities for $p(x)$, subject to (4.1.1), are

$$p(x) = 2x^3 + dx + e$$

or

$$p(x) = 2x^3 + 2x^2 + dx + e,$$

where d and e are arbitrary integers.   //

Note that in Example 81, there are two additional unknowns (parameters) other than x in $p(x)$. These parameters are d and e and both of them can have only integer values. Frequently polynomials have parameters (but not additional expressions involving x) as coefficients. Some parameters give a geometric interpretation as the polynomial which depends on a parameter can be thought of as a geometric family of polynomials. For example, in $q(x) = x^2 + a$ the parameter a is the value $q(0)$. In other words, a represents the point where this polynomial intercepts the y-axis (*i.e.* at $(0, a)$). For our solution to Example 81, e is the value of $p(x)$ when $p(x)$ intersects the y-axis.

If we ask what the possibilities for a polynomial are or ask you to write the polynomial $p(x)$ with as few parameters as possible, it means to find the coefficients in a way that relies on the smallest number of parameters as possible. As a reminder, a non-zero CONSTANT POLYNOMIAL has degree zero, a LINEAR POLYNOMIAL is a polynomial of degree 1, a QUADRATIC has degree 2, a CUBIC has degree 3, a QUARTIC has degree 4, and a QUINTIC has degree 5.

Exercise 44. If $p(x) = a(x^2 + 2x - 1) + b(x^2 + 3) + x - 5$ is a constant polynomial, then write the expression for $p(x)$ in as few parameters as possible.

Exercise 45. Let $p(x) = a(x^3 + x^2) + b(x^3 + x - 2) + x^2 + ax + 2$ be a linear polynomial. Write $p(x)$ with as few parameters as possible. What is $p(x)$ called?

Exercise 46. Let $p(x) = b_0 x^n + b_1 x^{n-1} + \cdots + b_{n-1} x + b_n$ where $b_0 \neq 0, b_i \in \mathbb{Z}$, $i = 0, 1, 2, \ldots, n$ and $n + |b_0| + |b_1| + |b_2| + \cdots + |b_n| = 2$. Find $p(x)$ if

A. $p(x)$ is a constant polynomial.
B. $p(x)$ is a linear polynomial.
C. $p(x)$ has degree n.

In Exercise 46, we asked you to solve a question in a way which illustrates the important problem-solving technique: *gaining insight by trying special cases.* Note that the coefficient of Exercise 46 are integers. After studying Exercise 46A and 46B, which are straightforward, Exercise 46C asks for the general case, which is now easier to solve because you've done the two special cases already. If, in the future, you are given a problem like Exercise 46C without 46A and 46B, try a few special cases ($n = 0$ or $n = 1$ or more) to gain insight before attempting to do the general problem for any n.

We now define addition and multiplication of polynomials in the usual way.

Definition. Let

$$p_1(x) = a_n x^n + a_{n-1} x^{n-1} \cdots + a_1 x + a_0$$

and

$$p_2(x) = b_m x^m + b_{m-1} x^{m-1} \cdots + b_1 x + b_0. \tag{4.1.2}$$

The SUM OF THE POLYNOMIALS $p_1(x)$ and $p_2(x)$, is, firstly, the polynomial $(p_1 + p_2)(x) = p_1(x) + p_2(x)$ of degree ℓ where ℓ is equal to the larger of the degrees m and n; secondly, for $p_1(x) + p_2(x)$, the coefficients are the same as the sum of the corresponding coefficients of $p_1(x)$ and $p_2(x)$. That is,

$$(p_1 + p_2)(x) = c_\ell x^\ell + c_{\ell-1} x^{\ell-1} + \cdots + c_1 x + c_0 \tag{4.1.3}$$

where $c_i = a_i + b_i$ for each i between 0 and ℓ. Therefore, the sum of two polynomials, $p_1(x)$ and $p_2(x)$, is defined as $(p_1 + p_2)(x) = p_1(x) + p_2(x)$.

There is a technical issue with Eq. (4.1.3) which is immediately discovered. It may be that the degrees of $p_1(x)$ and $p_2(x)$ are different, say $n < m$. In that case, there are no coefficients $a_{n+1}, a_{n+2}, \ldots a_m$. To make sense of the formula (4.1.3) it is then tacitly assumed that $a_{n+1} = a_{n+2} = \cdots = a_m = 0$. Thus $(p_1 + p_2)(x) = p_1(x) + p_2(x) = b_m x^m + \cdots + b_{n+1} x^{n+1} + (a_n + b_n) x^n + \cdots + (a_1 + b_1) x + (a_0 + b_0)$. Another technical point occurs when the leading coefficient of $p_1(x)$ is negative of the leading coefficient of $p_2(x)$ if p_1 and p_2 have the same degree. The technicality can also come in the product of polynomials. Note Eq. (4.1.4).

Definition. Given polynomials $p_1(x)$ and $p_2(x)$ of (4.1.2), the PRODUCT OF THE POLYNOMIALS is the polynomial of degree $n + m$ given by

$$(p_1 p_2)(x) = p_1(x)p_2(x) = a_n b_m x^{n+m}$$

$$+ \cdots + (a_2 b_0 + a_1 b_1 + a_0 b_2)x^2 + (a_1 b_0 + a_0 b_1)x + a_0 b_0.$$

In other words,

$$p_1(x) \cdot p_2(x) = c_{n+m} x^{n+m} + \cdots + c_1 x + c_0,$$

where the ith coefficient, c_i, is given by the formula

$$c_i = a_i b_0 + a_{i-1} b_1 + \cdots + a_0 b_i, \quad i = 0, \ldots, n+m. \tag{4.1.4}$$

Thus, the product of two polynomials, $p_1(x)$ and $p_2(x)$ is defined by $(p_1 \cdot p_2)(x) = p_1(x)p_2(x)$.

To help to remember Eq. (4.1.4), the sum of the two subscripts on each term in the right hand side is always i. Furthermore, Eq. (4.1.4) is exactly the formula for the conventional method of multiplying polynomials from middle or high school.

Example 82. The formal definitions of addition and multiplication of polynomials are described in Eqs. (4.1.3) and (4.1.4). Here are two polynomials, $f(x) = 3x^2 + 4x + 1$ and $g(x) = 2x^2 - x + 4$ that will be used in parts (a), (b) and (c).

(a). To compute $f(x) + g(x)$, with our proceeding notations, $a_2 = 3, a_1 = 4$ and $a_0 = 1$ whereas $b_2 = 2, b_1 = -1$ and $b_0 = 4$. Hence

$$f(x) + g(x) = (3x^2 + 4x + 1) + (2x^2 - x + 4)$$

$$= (3 + 2)x^2 + (4 - 1)x + (1 + 4)$$

$$= 5x^2 + 3x + 5.$$

(b). Similarly, for product of $f(x)$ and $g(x)$:

$$f(x) \cdot g(x) = c_4 x^4 + c_3 x^3 + c_2 x^2 + c_1 x + c_0, \tag{4.1.5}$$

where

$c_4 = a_4 b_0 + a_3 b_1 + a_2 b_2 + a_1 b_3 + a_0 b_4 = 0 \cdot 4 + 0 \cdot (-1) + 4 \cdot 0 + 1 \cdot 0 = 6$

$c_3 = a_3 b_0 + a_2 b_1 + a_1 b_2 + a_0 b_3 = 0 \cdot 4 + 3 \cdot (-1) + 4 \cdot 4 + 1 \cdot 0 = 5$

$c_2 = a_2 b_0 + a_1 b_1 + a_0 b_2 = 3 \cdot 4 + 4 \cdot (-1) + 1 \cdot 2 = 10$

$c_1 = a_1 b_0 + a_0 b_1 = 4 \cdot 4 + 1 \cdot (-1) = 15$

$c_0 = a_0 b_0 = 1 \cdot 4 = 4.$

Thus, $f(x)g(x) = 6x^4 + 5x^3 + 10x^2 + 15x + 4.$

(c). Alternatively, to compute $f(x) \cdot g(x) = (3x^2 + 4x + 1)(2x^2 - x + 4)$ we could multiply and collect like terms:

$$
\begin{aligned}
(3x^2 + 4x + 1)(2x^2 - x + 4) &= 3x^2(2x^2 - x + 4) \\
&\quad + 4x(2x^2 - x + 4) \\
&\quad + 1(2x^2 - x + 4) \\
&= 6x^4 + 5x^3 + 10x^2 + 15x + 4.
\end{aligned}
$$

We could also write vertically (by expanding the product)

$$
\begin{array}{r}
3x^2 + 4x + 1 \\
\times \quad\quad 2x^2 - \ x + 4 \\
\hline
12x^2 + 16x + 4 \\
- 3x^3 - 4x^2 - \quad x \\
6x^4 + 8x^3 + \ 2x^2 \\
\hline
6x^4 + 5x^3 + 10x^2 + 15x + 4
\end{array}
\tag{4.1.6}
$$

$$\uparrow$$

Note that the coefficients displayed vertically between the bars in this example are exactly the products in (4.1.4). For example, the coefficient of the quadratic term is

$$
c_2 = a_2 b_0 + a_1 b_1 + a_0 b_1 = 12 + 4(-1) + 1 \cdot 2 = 10,
$$

which comes from the coefficients in the column above the arrow.

Exercise 47. Compute the following additions and multiplications.

A. What polynomial is the sum of $x^3 - x^2 + 72x - 5$ and $x^3 - 4x + 1$?
B. Add $x^4 + 2x^2 - 1$ and $3x^3 + 4x^2 - x + 5$.
C. What polynomial is the product $(x^3 - x^2 + 72x - 5)(x^3 - 4x + 1)$?
D. What is $(x^4 + 2x^2 - 1)(3x^3 + 4x^2 - x + 5)$?

Exercise 48. Find the coefficients of x^7 and x^{10} respectively for the product $(x^5 + 2x^4 - x^3 + x^2 - 4x + 5)(x^6 + 7x^5 - 4x^3 + 2x + 1)$.

Example 83. Let $p(x) = (x^2 - ax + b)(ax^2 + x - b)$ be given where $a, b \in \mathbb{R}$. If, for $p(x)$, the coefficient of x^2 is 1 and the coefficient of x is 9, find the values of a and b.

Solution. Since $p(x) = (x^2 - ax + b)(ax^2 + x - b)$, the quadratic term is $(-b - a + ab)x^2$ and the linear term is $(ab + b)x$. So we gain the system of equations:

$$-b - a + ab = 1. \tag{4.1.7}$$

$$ab + b = 9. \tag{4.1.8}$$

Subtracting these two equations yields

$$-2b - a = -8 \text{ so } a = 8 - 2b.$$

Substituting the last equation into (4.1.7), we obtain

$$2b^2 - 9b + 9 = (2b - 3)(b - 3) = 0,$$

and so $b = \frac{3}{2}$ or 3.

There are two cases. If $b = \frac{3}{2}$ then $a = 8 - 2 \cdot \frac{3}{2} = 5$ and if $b = 3$ then $a = 8 - 2 \cdot 3 = 2$. So the two solutions are $a = 5, b = \frac{3}{2}$ and $a = 2, b = 3$.   //

Example 84. If $5x^2 = a(x^2 + x + 2) + (bx + c)(x + 1)$, what are the possible values for the real numbers a, b, and c?

Solution. Since $5x^2 = (a + b)x^2 + (a + b + c)x + (2a + c)$, by the definition of equality for polynomials we obtain the system of equations

$$\begin{cases} a + b = 5 \\ a + b + c = 0 \\ 2a + c = 0. \end{cases}$$

Solving these three equations in three unknowns yields only one solution

$$c = -5, a = \frac{5}{2}, \text{ and } b = \frac{5}{2}.$$

//

Example 85. (Version 1) If

$$a(x - 2)(x - 3) + b(x - 1)(x - 2) + c(x - 3)(x - 1) = 3x - 4, \tag{4.1.9}$$

then what are the real numbers a, b, and c?

Solution. By collecting like terms we can rewrite (4.1.9) as:

$$(a + b + c)x^2 + (-5a - 3b - 4c)x + (6a + 2b + 3c) = 3x - 4.$$

Equating coefficients of the two sides of the equality gives the following three equations:

$$\begin{cases} a + b + c = 0 \\ -5a - 3b - 4c = 3 \\ 6a + 2b + 3c = -4. \end{cases}$$

Solving the above system of equations, we obtain $a = -\frac{1}{2}, b = \frac{5}{2}, c = -2$ as the only solution. These values satisfy Eq. (4.1.9).   //

Often there are multiple ways to solve a problem where one of which is faster than the others. Generally, the faster approach requires a conceptual insight into the mechanics of the problem. In the case of Example 85, the insight is the interpretation of the polynomial of (4.1.9) as a function and recognizing that when $x = 1$ or $x = 2$ or $x = 3$ the arithmetic becomes much easier.

Example 86. (Version 2) Solve Example 85 in another way.

Solution. Since Eq. (4.1.9) is true for any $x \in \mathbb{R}$, consider the case where $x = 1, x = 2$, or $x = 3$. These three values of x infiltrate a number of zeros of the Eq. (4.1.9) that make things much easier.
  If $x = 1$, then

$$a(-1)(-2) = 3 \cdot 1 - 4 \text{ so that } 2a = -1 \text{ or } a = -\frac{1}{2}.$$

On the other hand, if $x = 2$, then

$$c(2 - 3)(2 - 1) = 3 \cdot 2 - 4 \text{ so that } -c = 2 \text{ or } c = -2.$$

And finally if $x = 3$, then

$$b(2)(1) = 9 - 4 \text{ so that } 2b = 5 \text{ or } b = \frac{5}{2}.$$

//

Why did the solution fall out so simply in the second version? The problem-solving strategy was to *think of a polynomial as a function* and notice that when $x = 1, 2$ or 3, one or two of the terms on the left hand side were zero. Said in other words, our *problem-solving strategy meant finding a different interpretation* (in this case as a function) and looking at the equation to see what values might simplify the arithmetic.
  It is true, of course, that any three values plugged into Eq. (4.1.9) would ultimately yield the result $a = -\frac{1}{2}, b = \frac{5}{2}$, and $c = -2$. However, the strategy to pick numbers judiciously made the arithmetic much easier and was the most advantageous approach.

The moral (which would be wise to follow in Exercise 49 below) is to *make sure you stop and think before starting a problem*. A few minutes of thinking about problem-solving strategies will result in much less work and frequently, additional insight.

Exercise 49. Assume that $x^4 - 8x^3 + 5x^2 - 30x + 8 = a + b(x-1) + c(x-1)$ $(x-2) + d(x-1)(x-2)(x-3) + e(x-1)(x-2)(x-3)(x-4)$. Find the real numbers a, b, c, d, and e or, if there is no solution, explain why.

Problem Set 4.1:

Through this problem set, a, b, c, d, and e are real numbers unless otherwise stated.

1. Simplify $x^2(4x^2 - 2x - 6) - (x-1)^2(4x^2 + x - 2)$.
2. If $p(x) = a(x^2 - x + 1) + b(x^2 + 2x) + 1$ is the zero polynomial, what are a and b?
3. Let $g(x) = r(x^3 - x^2) + s(x^2 + 2x) + 3x^2 + tx - 1$ with $r, s, t \in \mathbb{R}$ be a linear polynomial. Write $g(x)$ with as few parameters as possible.
4. If $h(x)$ is a quadratic polynomial, write $h(x)$ with the fewest parameters possible.
5. (a). If $f(x) = ax^3 + (b-1)x^2 + (c+3)x - 1$ and $g(x) = 3x^2 + (a-4)x + (b-a+c)$, find a, b, and c so that $f(x) = g(x)$ (if there is a solution).
   (b). Find a, b, and c (if possible) so that the equation $3x^3 - 4x^2 + 5x - c = (x^2 - 2x + 3)(ax + b)$ is valid.
6. Find, if possible, integers a, b and c so that $(a+b+c)x^3 + (|a|+|b|)x + c = 0$.
7. Find, if possible, integers a, b, and c such that $(a+b+c)x^3 + |a-b|x^2 + (|a|+|b|)x + c = 0$.
8. Find, if possible, real numbers a, b, and c so that $3x^3 - 4x^2 + 6x - c = (x^2 - 2x + 3)(ax + b)$.
9. Find the coefficients of x^9 and x^{11} in the polynomial $q(x)$ where

$$q(x) = (x^7 + 2x^6 - 5x^3 + x^2)(7x^{10} - 33x^{11} + 5x^2 - 7x^5 + 23x^9)$$

10. Let $3x^2 + 4x - 5 = a(x+1)(x-1) + b(x-1)(x-2) - (x-2)(x+1)$. Find the value of a and b.
11. If $(x-1)(x-2)(x+3) = x^3 + ax^2 + bx + c$, find the value of $a + b + c$.
12. If $(x-1)(x+2)(x-3)(x+4) = ax^4 + bx^3 + cx^2 + dx + e$, then

   (a) what is e?
   (b) What is $a + b + c + d$?

13. Let $2(x+3)^4 - 15(x+3)^3 + 31(x+3)^2 - 8(x+3) + 8 = ax^4 + bx^3 + cx^2 + dx + e$. Find the values of a, b, c, d, and e.
14. If $54x^3 + 27x^2 + 42x + 32 = a(3x+2)^3 + b(3x+2)^2 + c(3x+2) + d$, find the value of $a + c$. (Hint: Find the value of $a + b + c + d$ first.)

15. How would you explain where formula (4.1.4) appears in your solution to Exercise 47A to an Algebra I class? (Hint: Write out the solution in a form similar to Example 82. Look at the "scaffolding" (4.1.6) for the explanation.)

16. Let $p(x) = b_0 x^n + b_1 x^{n-1} + \cdots + b_{n-1} x + b_n$ where $p(x)$ has degree n. If

$$|b_0| + |b_1| + \cdots + |b_n| = n$$

find all possible polynomials $p(x)$ if

(a). $p(x)$ is a constant polynomial.
(b). $p(x)$ is a linear polynomial.
(c). $p(x)$ is a quadratic polynomial.
(d). $p(x)$ is a polynomial of degree n.

17. Let $p(x) = (x + a)(x + b)(x + c)$ and $q(x) = (x - a)(x + b)(x + c)$ where $a, b, c \in \mathbb{R}$. Assume

1. the coefficient of x^2 of $p(x)$ is 0
2. the coefficient of x of $q(x)$ is 0
3. the coefficient of x of $p(x)$ is equal to the coefficient of x^2 of $q(x)$.

Prove that $a = 1$ and $bc = b + c$.

18. Let $(x^2 - 5x + 4)(x^2 - 7x + 5) = ax^4 + bx^3 + cx^2 + dx + e$. Find the value of $a + b + c + d + e$.

19. If the coefficient of x^3 in polynomial $(3x^2 - 2x + 6)(x^3 + ax^2 + 8)$ is -28, then what is the value of a? (Answer: $a = 17$)

4.2 Divisibility, Quotients and Remainders of Polynomials

Overview Our aim for the next sections is to show how properties of the integers and those of the real numbers are similar to properties of the set of all polynomials in one variable. In this section, we see that formally, the Division Algorithm for Polynomials is similar to the Division Algorithm for Integers of Sect. 1.1. The definitions, theorems, and even the proofs in this section are very similar to corresponding statements in Chap. 1. Our viewpoint will be strengthened in Sect. 5.3 when we discuss the GCD and LCM for two polynomials.

Recall the Division Algorithm in $\mathbb{Z}$: If $a, b \in \mathbb{Z}$, with $b > 0$, then there exists unique integers q and r such that

$$a = qb + r, \quad 0 \le r < b.$$

The algorithm by which q (the *quotient*) and r (the *remainder*) are found is just long division (or successive subtraction). Similarly, there is the Division Algorithm for Polynomials, which also can be proven by successive subtraction. Note that we cannot tell when one polynomial is greater than another, so instead we compare their degrees.

Theorem 15 (Division Algorithm for Polynomials). *Let $f(x)$ be a polynomial and $g(x)$ be a non-zero polynomial. Then there exist unique polynomials $q(x)$ and $r(x)$ such that*

$$f(x) = q(x) \cdot g(x) + r(x),$$

where either $r(x) = 0$ or $\deg r(x) < \deg g(x)$. We call the polynomial $q(x)$ the QUOTIENT, *$r(x)$ the* REMAINDER, *and $g(x)$ the* DIVISOR *of $f(x)$. We also say that $f(x)$ when divided by $g(x)$ has remainder $r(x)$ and quotient $q(x)$.*

Example 87. Let $f(x) = 2x^3 + 4x^2 + x - 2$ and $g(x) = 2x^2 + x - 1$ be the divisor of $f(x)$. Find the quotient and remainder of $f(x)$ when divided by $g(x)$.

Solution. We divide $f(x) = 2x^3 + 4x^2 + x - 2$ by $g(x) = 2x^2 + x - 1$ using long division.

$$
\begin{array}{r}
x + \frac{3}{2} \\
2x^2 + x - 1\overline{)2x^3 + 4x^2 + x - 2} \\
\underline{2x^3 + x^2 - x} \quad \longleftarrow \\
3x^2 + 2x - 2 \\
\underline{3x^2 + \frac{3}{2}x - \frac{3}{2}} \\
\frac{1}{2}x - \frac{1}{2}
\end{array}
$$

$$(4.2.1)$$

Hence

$$2x^3 + 4x^2 + x - 2 = \left(x + \frac{3}{2}\right)(2x^2 + x - 1) + \left(\frac{1}{2}x - \frac{1}{2}\right). \qquad (4.2.2)$$

How do we interpret Eq. (4.2.2) in terms of the Division Algorithm for Polynomials? The Eq. (4.2.2) says that

$$f(x) = \underbrace{\left(x + \frac{3}{2}\right)}_{\text{quotient, } q(x)} g(x) + \underbrace{\left(\frac{1}{2}x - \frac{1}{2}\right)}_{\text{remainder, } r(x)} \qquad (4.2.3)$$

so the quotient is $q(x) = x + \frac{3}{2}$ and the remainder is $r(x) = \frac{1}{2}x - \frac{1}{2}$. The remainder $r(x)$ has degree 1, which is less than $\deg g(x) = 2$.   //

Similarity of Division Algorithm for Polynomials and Division Algorithm for Integers. There is something more going on in the solution of Example 87. Looking at the long division, we see that (just as for integers) we are doing successive subtractions. For example, the line directly under $f(x) = 2x^3 + 4x^2 + x - 2$

in the division (4.2.1) is $2x^3 + x^2 - x = xg(x)$. Following the process for the case of integers, we subtract the line directly under $f(x)$ from $f(x)$ to obtain the next line, $3x^2 + 2x - 2$. We now explain why in the next paragraph. Just as in the integer case, the degree is lowered by 1!

What follows in this paragraph is an intuitive explanation of why there is an analogy between long division of integers and the division of polynomials. We will use $f(x)$ and $g(x)$ from Example 87 and see that the first step in the division of $f(x)$ by $g(x)$ gives a polynomial $(2x^3 + x^2 - x)$ at the arrow in Eq. (4.2.1) whose degree is 2 (one less than the degree of $f(x)$). This is similar to starting with a four digit number, say $2314 = 2 \cdot 10^3 + 3 \cdot 10^2 + 1 \cdot 10 + 4$, and getting a first step (at the arrow below to a three digit number) or a step which takes a "10^3 number" to a "10^2 number". In a numerical example, dividing 2314 by 11 is written in long division as

$$
\begin{array}{r}
210 \\
11\,)\,\overline{2314} \\
\underline{2200} \\
114 \leftarrow \\
\underline{110} \\
4
\end{array}
$$

The number 114, at the arrow represents reducing the degree by one—going from a $10^3 = 1000$ term (2314) to a $10^2 = 100$ term (114). Thus, in the first step, for integers in base 10, we are going from x^3 to x^2 where $x = 10$. (The second step reduces from x^2 to x^1 when $x = 10$.) This intuitive motivation should be helpful to you but note that not every division of an integer by another integer "reduces the power of 10 by one". For example, divide 99 by 22 or 9 by 2.

Long division of polynomials is exactly successive subtractions (just as for integers). In Example 87, we have "subtracted $g(x)$ a total of x times" from the $f(x)$ and then subtracted $\frac{3}{2}g(x)$ from what is left. The product $xg(x)$ has degree one more than $\frac{3}{2}g(x)$. If the original polynomial $f(x)$ has degree n, then there will be at most n subtractions to perform. (In fact, having $(n - 1)$ subtractions can only occur if $\deg g(x) = 1$. Usually, there are fewer calculations.)

We will now use the Division Algorithm for Polynomials in which there are different divisors of the given polynomials $f(x) = ax^2 + bx - 4$.

Example 88. Let $f(x) = ax^2 + bx - 4$ be divided by $x + 1$ and $x - 1$ with respective remainders 3 and 1. What is the remainder of $f(x)$ when divided by $x + 2$?

Solution. There are three different quotients $q_1(x)$, $q_2(x)$ and $q_3(x)$ starting with $f(x)$. The statements of the first sentence of the example are given in part (a) and part (b). Part (c) uses the quotient $q_3(x)$ with a remainder r.

$$f(x) = ax^2 + bx - 4 = q_1(x)\underbrace{(x+1)}_{\text{divisor}} + \underbrace{3}_{\text{remainder}} \qquad (a)$$

$$f(x) = q_2(x)(x-1) + 1 \qquad (b)$$

$$f(x) = q_3(x)(x+2) + r. \qquad (c)$$

Then we obtain

$$f(-1) = a(-1)^2 + b(-1) - 4 = 3,$$

$$f(1) = a(1)^2 + b(1) - 4 = 1,$$

$$f(-2) = a(-2)^2 + b(-2) - 4 = r.$$

i.e.

$$a - b - 4 = 3, \quad \text{so that } a - b = 7,$$

$$a + b - 4 = 1, \quad \text{so that } a + b = 5,$$

$$\text{and} \quad 4a - 2b - 4 = r.$$

Solving this system of the first two equations, we have $a = 6, b = -1$ and $r = 4a - 2b - 4 = 4(6) - 2(-1) - 4 = 22.$ //

Example 89. Let $f(x) = ax^2 + bx + c$ be divided by $x + 1$ and $2x - 1$ with respective remainders 1 and -2. What is the remainder when $f(x)$ is divided by $2x^2 + x - 1$?

Solution. When $f(x)$ has a divisor $(2x^2 + x - 1)$ which is also a quadratic, the remainder must be linear or constant, say $ax + b$ where $a, b \in \mathbb{R}$. Let $f(x) = q(x)(2x^2 + x - 1) + ax + b$ where $q(x)$ is the quotient polynomial. Then

$$f(-1) = -a + b = 1$$

$$f(1/2) = (a/2) + b = -2$$

Solving this system we find $a = -2$ and $b = -1$. Thus the remainder is $-2x - 1$. //

Example 90. Let $f(x)$ be divided by $x+2$ and x^2-x-2 with respective remainders 3 and $5x + 1$. What is the remainder when $f(x)$ is divided by $x^2 - 4$?

Solution. $f(x) = q(x)(x^2 - 4) + ax + b$ where $q(x)$ is a quotient. By evaluating $x = -2$ and $x = 2$, we see first that

$$f(-2) = -2a + b = 3. \qquad (4.2.4)$$

With $q_1(x)$ as a quotient of $x^2 - x - 2$, $f(x) = q_1(x)(x^2 - x - 2) + 5x + 1$, we have

$$f(2) = 2a + b = 11 \qquad\qquad (4.2.5)$$

Solving (4.2.4) and (4.2.5), we obtain $a = 2$ and $b = 7$. Hence the remainder is $2x + 7$.   //

Example 91. What is the remainder when $x + x^9 + x^{25} + x^{49} + x^{81}$ is divided by $x^3 - x$? (See Honsberger [5])

Solution. We present a different and very easy way to find the remainder other than [5]. Let $x + x^9 + x^{25} + x^{49} + x^{81} = q(x)(x^3 - x) + ax^2 + bx + c$, where $ax^2 + bx + c$ is the remainder we want to find. (As our divisor is of degree 3, we know that our remainder should be at most degree 2.) Since the roots of $x^3 - x$ are $x = 0$ and $x = \pm 1$, we can now find the remainder in the form of $ax^2 + bx + c$ and a quotient $q(x)$.

Now we can use the roots of $x^3 - x$ to isolate our unknowns a, b, c. Letting $x = 0$ immediately results in $0 = c$. Letting $x = 1$ gives $5 = a + b$, and letting $x = -1$ gives $-5 = a - b$. Solving this simple system gives us that $a = 0$ and $b = 5$. Hence the remainder is $5x$.   //

Example 92. Let $p(x)$ be a polynomial with the property that if $p(x)$ is divided by $(x - 2)^2$ a remainder of $56x - 42$ is obtained. If $p(x)$ is divided by $(2x - 1)$ a remainder of 5 is obtained. What is the remainder if $p(x)$ is divided by $(x - 2)^2(2x - 1)$?

Solution. The minimum degree of $p(x)$ is 3. From the first sentence we know that there is a polynomial $q_1(x)$ with

$$p(x) = q_1(x)(x - 2)^2 + (56x - 42).$$

Using the Division Algorithm for $q_1(x)$ with divisor $(2x - 1)$ gives a polynomial $q(x)$ and a real number a such that $q_1(x) = q(x)(2x - 1) + a$. Thus

$$p(x) = (x - 2)^2(2x - 1)q(x) + a(x - 2)^2 + 56x - 42. \qquad (4.2.6)$$

We now let $x = \frac{1}{2}$. Then $p\left(\frac{1}{2}\right) = \frac{9a}{4} + 28 - 42 = 5$ so that $a = \frac{76}{9}$. Substituting $a = \frac{76}{9}$ into Eq. (4.2.6) gives a remainder of $\frac{76}{9}(x-2)^2 + 56x - 42 = \frac{1}{9}(76x^2 + 200x - 74)$. //

Exercise 50. If a given polynomial $p(x)$ is divided by $x^2 + x + 1$ a remainder of 3 is obtained. If $p(x)$ is divided by $x + 5$, a remainder of 7 is obtained. What is the remainder, r, when $p(x)$ is divided by $(x^2 + x + 1)(x + 5)$?

Exercise 51. What is the remainder and quotient when one divides the polynomial $p(x) = 4x^4 + 2x^3 - 4x^2 + 6x - 1$ by $g(x) = 2x^2 - x - 1$? (Answer for remainder is $8x - 1$.)

Problem Set 4.2:

1. Let $f(x) = 4x^3 + 8x^2 + x - 1$ and $g(x) = x + 2$. Find the quotient and remainder of $f(x)$ by $g(x)$ by doing long division. Compare your answer to Example 100.
2. Let $f(x) = 9x^3 + 4x^2 - 5x - 2$. Find the quotient and remainder of $f(x)$ on division by $2x - 1$ by using long division. Compare your answer to Example 101.
3. Find the quotient and remainder if the first polynomial is divided by the second. Check your answer by multiplying the expressions.

   (a) $x^3 - 2x^2 + 3x + 1$ by $x^2 - x + 2$
   (b) $4x^3 + x^2 - 12x - 3$ by $x^2 - 3$

4. Given a polynomial $p(x)$, if $p(x)$ is divided by $x - 1$ a remainder of 8 is obtained. If $p(x)$ is divided by $x^2 + x + 1$ a remainder of $7x + 16$ is obtained. What is the remainder if $p(x)$ is divided by $(x - 1)(x^2 + x + 1)$?
5. A polynomial $p(x)$, with degree ≥ 3 is given. If $p(x)$ is divided by $(x - 1)$, $(x - 2)$, and $(x - 3)$ remainders of 3, 7, and 13 are obtained, respectively. Find the remainder if $p(x)$ is divided by $(x - 1)(x - 2)(x - 3)$.
6. The polynomial $p(x)$ with degree ≥ 3 has the following two properties. If $p(x)$ is divided by $(x - 1)(x - 2)$ a remainder of $(20x - 6)$ is obtained. If $p(x)$ is divided by $(x - 2)(x - 3)$ a remainder of $32x - 30$ is obtained. Find the remainder when $p(x)$ is divided by $(x - 1)(x - 2)(x - 3)$.

4.3 The Remainder Theorem

Overview The root of solution of a polynomial equation has a very careful relationship to the Remainder Theorem. Let $f(x) = a_n x^n + a_{n-1} x^{n-1} + \cdots + a_1 x + a_0$ be a polynomial of degree n. For any real number c, we define $f(c)$ by the usual formula,

$$f(c) = a_n c^n + a_{n-1} c^{n-1} + \cdots + a_1 c + a_0.$$

That is, $f(c)$ is obtained when the constant c is substituted for x.

Definition. If $f(x)$ is a polynomial and $c \in \mathbb{R}$ then c is a ROOT of $f(x)$ if $f(c) = 0$. We also say that c is a SOLUTION of $f(x) = 0$ if $f(c) = 0$.

One of the conclusions of the Division Algorithm (Theorem 15) is that for c a real number, regardless of what polynomial is divided by $x - c$, the degree of the remainder must be less than one. Thus, the remainder is a real number r rather than a polynomial of degree one or higher. If $q(x)$ is the quotient of $f(x)$ upon division by $x - c$ then we have

$$f(x) = q(x)(x - c) + r.$$

Evaluating this equation at $x = c$ gives the following Remainder Theorem.

Theorem 16 (Remainder Theorem). *If the polynomial $f(x)$ is divided by $x - c$, then the remainder is $r = f(c)$.*

Corollary 7. *If the polynomial $f(x)$ is divided by $ax - b$, then the remainder is $f\left(\frac{b}{a}\right)$.*

Throughout this text we have used the usual vocabulary of proof technique. A COROLLARY means an easy consequence of the theorem which precedes it. Thus, in Exercise 52 below, it is natural to use Theorem 16 to prove the Corollary 7.

Exercise 52. Prove Corollary 7

Example 93. If $f(x) = x^4 + 5x^3 + 4x^2 + x - 5$ is divided by $x - 4$, find the remainder.

Solution. We can compute the remainder in two different ways: directly dividing $f(x)$ by $x - 4$ or using the Remainder Theorem. We'll do both so that you'll see what is involved in the two methods and can make a choice of method in future problems.

(a) Using long division by $x - 4$ and, after much arithmetic, we see

$$f(x) = (x^3 + 9x^2 + 40x + 161)(x - 4) + 639.$$

Thus, the remainder of $f(x)$ on division by $x - 4$ is 639 and so the remainder is 639.

(b) The other way to do this problem is to simply substitute $x = 4$ in $f(x)$ to obtain

$$f(4) = (4)^4 + 5(4^3) + 4(4^2) + 4 - 5 = 639.$$

$/\!/$

Example 94. Find the remainder if $f(x) = x^{15} + 1$ is divided by $x - 1$.

Solution. By the Remainder Theorem, the remainder is $f(1)$, hence the remainder is $f(1) = 1^{15} + 1 = 2$. $/\!/$

Remark. The point is that we did not have to use either long division or synthetic division (see Sect. 4.4) in this problem.

Example 95. When $f(x) = 2x^3 - x^2 + ax - 9$ is divided by $2x - 3$ the remainder is 0. What is the value of a

Solution. By the corollary of the Remainder Theorem, we have the remainder $f\left(\frac{3}{2}\right)$ which is given to be 0. Thus, if $\frac{3}{2}$ is a root of $f(x)$, then

$$2\left(\frac{3}{2}\right)^3 - \left(\frac{3}{2}\right)^2 + a\left(\frac{3}{2}\right) - 9 = 0.$$

After simplifying the equation we see that

$$\frac{27}{4} - \frac{9}{4} + \frac{3}{2}a - 9 = 0 \text{ or } 27 - 9 + 6a - 36 = 0.$$

Hence $a = 3$. We have shown that if there is a value of a such that $f(x)$ has $\frac{3}{2}$ as a root, it must be $a = 3$. We check our solution by calculating $f\left(\frac{3}{2}\right) = 0$ for $a = 3$.
//

Example 96. Find the remainder if $f(x) = x^5 - 4x^4 + x^3 + 9x^2 + 10x - 6$ is divided by $x - 1 - \sqrt{2}$.

Solution. By the Remainder Theorem, we know that the remainder r must satisfy $r = f(1+\sqrt{2})$. One way to do that is to plug $x = 1+\sqrt{2}$ into the equation for $f(x)$ but that gives a very tedious computation so we will look for another approach (a mathematician might say "it gets ugly fast", although this approach will ultimately lead to a correct answer). Surely there must be a different way to do the problem that requires less computation!

Let's **start again**, using some algebraic intuition to simplify the calculation. Let $x = 1 + \sqrt{2}$ so that $x - 1 = \sqrt{2}$. Thus $(x - 1)^2 = 2$ and $x^2 - 2x + 1 = 2$, so that $x^2 - 2x - 1 = 0$. Motivated by that computation, we now use the division algorithm to divide $f(x)$ by $x^2 - 2x - 1$ (instead of by $x - (1 - \sqrt{2})$), the remainder of $f(x)$ divided by $x^2 - 2x - 1$ will be a linear polynomial $r(x)$ since the divisor is quadratic.

$$x^5 - 4x^4 + x^3 + 9x^2 + 10x - 6 = q(x) \cdot (x^2 - 2x - 1) + r(x) \qquad (4.3.1)$$

We'll see why this approach is useful right after (4.3.2). Finding the quotient and remainder can easily be done by long division.

$$
\begin{array}{r}
x^3 - 2x^2 - 2x + 3 \\
x^2 - 2x - 1 \overline{)\, x^5 - 4x^4 + x^3 + 9x^2 + 10x - 6} \\
\underline{x^5 - 2x^4 - x^3} \\
-2x^4 + 2x^3 + 9x^2 \\
\underline{-2x^4 + 4x^3 + 2x^2} \\
-2x^3 + 7x^2 + 10x
\end{array}
$$

$$\underline{-2x^3 + 4x^2 + 2x}$$

$$3x^2 + 8x - 6$$

$$\underline{3x^2 - 6x - 3}$$

$$14x - 3$$

Hence for $f(x)$ divided by $x^2 - 2x - 1$, the quotient is $q(x) = x^3 - 2x^2 - 2x + 3$ and the remainder is $r(x) = 14x - 3$. This means that

$$f(x) = x^5 - 4x^4 + x^3 + 9x^2 + 10x - 6$$

$$= \underbrace{(x^3 - 2x^2 - 2x + 3)}_{q(x)} \underbrace{(x^2 - 2x - 1)}_{\text{divisor}} + \underbrace{(14x - 3)}_{r(x)}$$

or writing $q(x) = x^3 - 2x^2 - 2x + 3$,

$$f(x) = q(x)(x^2 - 2x - 1) + (14x - 3). \tag{4.3.2}$$

The key here is that we have found a form of Eq. (4.3.1) which is very easy to evaluate when $x = 1 + \sqrt{2}$. *The point is that $q(1 + \sqrt{2})$ need not be computed. Rather we notice that $q(x)$ is multiplied by $(x^2 - 2x - 1)$ which has $x = 1 + \sqrt{2}$ as a root of $f(x)$ and hence, using Eq. (4.3.2), its product with $q(1 + \sqrt{2})$ is also zero.* To finish the example,

$$f(1 + \sqrt{2}) = q(1 + \sqrt{2})0 + 14(1 + \sqrt{2}) - 3 = 11 + 14\sqrt{2}. \tag{4.3.3}$$

$$//$$

The solution to Example 96 is a case of *problem-solving by infiltrating zeros.* The approach is to involve the division algorithm in a way that greatly simplifies the arithmetic through multiplication by 0 (see Eq. (4.3.3)). Some thought shows that a linear term will not help much but "getting rid of the square root" might (and does). That approach requires squaring and so we end up using a quadratic polynomial $x^2 - 2x - 1$ as a divisor of $f(x)$. The key, of course, is that $x^2 - 2x - 1$ is zero when $x = 1 + \sqrt{2}$. An approach which seems to look like it will make things more complicated (going to a quadratic instead of using a linear polynomial) in fact makes the problem easier!

Remark. Since $1 - \sqrt{2}$ is also a root of $f(x) = x^2 - 2x - 1 = 0$, we have $f(1 - \sqrt{2}) = q(x) \cdot 0 + 14 \cdot (1 - \sqrt{2}) - 3 = 11 - 14\sqrt{2}$.

Exercise 53. Find the remainder if $x^4 - 3x^3 - 2x^2 + 1$ is divided by $(x - \frac{-1+\sqrt{5}}{2})$. (Answer: $\frac{15-7\sqrt{5}}{2}$.)

A final remark We can use the remainder theorem to give another proof of tests for divisibility (see Sect. 1.1).

Proof. Let $f(x) = a_n x^n + a_{n-1} x^{n-1} + \cdots + a_1 x + a_0$ be a polynomial of degree n with coefficients in $\mathbb{Z}$ and $0 \leq a_i \leq 9\,(i = 0, 1, \cdots, n)$, $a_n \neq 0$. If $f(x)$ is divided by $x - 1$, then the remainder theorem gives

$$a_n x^n + \cdots + a_1 x + a_0 = q(x)(x-1) + (a_n + a_{n-1} + \cdots a_1 + a_0).$$

Let $x = 10$, we have

$$a_n 10^n + a_{n-1} 10^{n-1} + \cdots + a_1 10 + a_0 = q(10) \cdot 9 + (a_n + \cdots + a_1 + a_0).$$

Thus, we have proved the divisibility test for 3 and 9. ∎

Theorem 17 (Divisibility Test 3/9). *Let m be represented as $m = a_n a_{n-1} \cdots a_2 a_1 a_0$. Then we have the following results:*

1. *3 divides m if and only if 3 divides $a_n + a_{n-1} \cdots + a_2 + a_1 + a_0$.*
2. *9 divides m if and only if 9 divides $a_n + a_{n-1} \cdots + a_2 + a_1 + a_0$.*

Exercise 54. By using the remainder theorem, prove the divisibility test for 11. (Hint: Consider $a_n x^n + a_{n-1} x^{n-1} + \cdots + a_1 x + a_0$ is divided by $x + 1$.)

Problem Set 4.3:

1. If $x^3 + 3x^2 - x - 6$ is divided by $2x - \sqrt{5}$, find the remainder by infiltrating zeros.
2. If $(x^3 + ax^2 + 3x - 4)$ is divided by $(x + 1)$ with remainder 2, find the value of a.
3. If $(x^3 + ax^2 + 5x + 1)$ is divided by $(2x - 1)$ with remainder 4, find the value of a.
4. If $(x^3 + ax^2 - 6x + 2)$ is divided by $(3x + 2)$ with remainder 10, find the value of a.
5. Let $p(x) = x^3 + ax^2 + bx - 2$. If $p(x)$, when divided by $x + 1$, has remainder -1 and $p(x)$, when divided by $x - 1$, has remainder 3, find the real numbers a and b.
6. Let $p(x) = x^3 + ax^2 + bx - 1$. If $p(x)$, when divided by $x + 2$, has remainder -11 and $p(x)$, when divided by $x - 2$, has remainder 17, find the real numbers a and b.
7. Let $p(x) = x^3 + ax^2 + bx - 6$. If $p(x)$, when divided by $x - 2$, has remainder 2 and $p(x)$, when divided by $x - 3$, has remainder 51, find the real numbers a and b.
8. Find the remainder if $p(x) = x^6 + 5x^5 - 10x^3 + 4x - 1$ is divided by $x - \sqrt[3]{3}$.
9. Find the remainder of

$$f(x) = x^5 + 4x^4 - 3x + 2$$

if it is divided by $x + 2 - \sqrt{3}$.

10. Find the remainder of

$$g(x) = x^4 - x^2 + 3x - 12$$

if it is divided by $x - 1 - \sqrt{5}$.

11. If $x^3 + 3x^2 - x - 6$ is divided by $x - (\sqrt{5} - 1)$, find the remainder.
12. If $4x^4 - 8x^3 - 15x^2 + 13x + 1$ is divided by $2x - (3 + 2\sqrt{2})$, find the remainder.
13. If $9x^4 - 54x^3 + 8x^2 + 8x - 1$ is divided by $x - (5 - 3\sqrt{2})$, find the remainder.
14. If $x^4 - 22x^2 - 48x + 2$ is divided by $x - (\sqrt{2} + \sqrt{3} - \sqrt{6})$, find the remainder.
15. If $x^4 + 11x^2 + 21x + 4$ is divided by $x - (\sqrt{3} + \sqrt{5} + \sqrt{10})$, find the remainder.
16. Find the remainder of $p(x) = x^6 + 4x^5 - x^3 + 1$ on division by $x - 1 + \sqrt[3]{2}$ by infiltrating zeros.
17. If $(x + 3)^n + 1$ is divided by $x + 1$, find the remainder (as a function of n).
18. Suppose that $p(x) = ax^2 + bx + c$ is a quadratic. If the remainder of $p(x)$ on division by $x - 7$ is 3 and the remainder of $p(x)$ on division by $x - 2$ is 4, show that a and b are relatively prime. (Hint: Use Proposition 5.)
19. Suppose that neither a or b is zero and $r \neq s$. If the remainder of $q(x) = ax^2 + bx + c$ on division by $x - r$ is d_1 and the remainder of $q(x)$ on division by $x - s$ is d_2, show that $GCD(a, b)$ divides $d_1 - d_2$.
20. Let $f(x) = x^2 - ax + b$. If $f(2) = 6$, $f(-1) = 3$, then what are real numbers a and b? (Ans: $a = 0$, $b = 2$.)
21. Let $f(x) = x^4 + px^2 + q(x) + a^2$. If $f(x)$ have two roots 1 and -1, then $f(x)$ has two other roots a and $-a$. Prove this result.

4.4   Synthetic Division

Overview This section is not necessary nor dependent on other sections, but it is quite interesting. Some calculation and notation can be saved by omitting powers of x in the long division of polynomials, although care must be taken to keep track of where each of the coefficients came from. This observation leads to the notational shortcut called *Horner's scheme* or *Synthetic Division* and involves division of any polynomial by another polynomial; however we will emphasize division by linear or quadratic polynomials. The method is taught in high school in the US under the name *Synthetic Division* and in other countries, such as Taiwan, at a similar grade level but usually called *Horner's Scheme*. We will use these phrases *Synthetic Division* and *Horner's Scheme* interchangeably; *Horner's scheme* is due to W.G. Horner (1786–1837) in about 1819 .

Example 97. Let $f(x) = 2x^3 + 4x^2 + x - 2$ and the devisor $g(x) = 2x^2 + x - 1$. Complete the polynomial division while omitting all powers of x.

Solution. Coordinating the polynomials $f(x)$ and $g(x)$ without any x terms is easily done in Table 4.1 below.

Table 4.1 The functions of Example 87

As polynomial		As coefficients	
$f(x) = 2x^3 + 4x^2 + x - 2$		$2 \quad +4 \quad +1 \quad -2$	
$g(x) = \qquad\quad 2x^2 + x - 1$		$2 \quad +1 \quad -1$	

Table 4.2 Equation (4.2.1) as coefficients

$$
\begin{array}{r}
1 + \frac{3}{2} \\
\hline
2 + 1 - 1 \overline{)\, 2 + 4 + 1 - 2} \\
2 + 1 - 1 \\
\hline
3 + 2 - 2 \\
3 + \frac{3}{2} - \frac{3}{2} \\
\hline
\frac{1}{2} - \frac{1}{2}
\end{array}
$$

Using only these coefficients, the polynomial division in Eq. (4.2.1) is represented by Table 4.2. Here the quotient appearing as $1 + \frac{3}{2}$ represents the polynomial $q(x) = 1x + \frac{3}{2}$ and the remainder appearing as $\frac{1}{2} - \frac{1}{2}$ represents the polynomial $r(x) = \frac{1}{2}x - \frac{1}{2}$. As expected, these quotient and remainder terms match those from Eq. (4.2.3).

//

Example 98. Let $h(x) = 2x^3 + x + 1$ and $g(x) = x^2 + x$. Find the quotient and remainder of $h(x)$ by $g(x)$ using only the coefficients in the polynomial division.

Solution. A word of caution is first in order as you construct a table similar to Tables 4.1 or 4.2. *Don't forget to include the zero coefficients!* Since, really, $h(x) = 2x^3 + x + 1$ and $g(x) = x^2 + x$, the table for this problem similar to what was done for Example 87 is:

$$
\begin{array}{r}
2 - 2 \\
\hline
1\ \ 1\ \ 0 \overline{)\, 2\ \ \ 0\ \ \ 1\ \ \ 1} \\
2\ \ \ 2\ \ \ 0 \\
\hline
-2\ \ \ 1\ \ \ 1 \\
2\ \ \ 2\ \ \ 0 \\
\hline
-1\ \ \ 1
\end{array}
$$

Thus the quotient is $q(x) = 2x - 2$ and the remainder is $r(x) = -x + 1$. You must make sure not to forget to put in all the terms (even if a coefficient is zero) when constructing a table.

//

Exercise 55. Use only the coefficients in long division to find the quotient $q(x)$ and the remainder $r(x)$ when the first polynomial is divided by the second. Before you start, decide what the degree of the quotient will be.

(a) $3x^3 + 4x^2 + 1$ by $x^2 + x + 1$
(b) $5x^5 + 4x^3 + 2x + x + 2$ by $x - 3$

In the event that the divisor is a monic linear polynomial $x - d$, then the divisor in Table 4.2 reduces to the form $1 - d$. At each stage of the coefficient division, the initial term of the product of the divisor with the quotient will mirror the quotient. In the next example, compare the initial term at each stage with the quotient term used to create it.

Example 99. Divide $f(x) = 2x^3 + 4x^2 + x - 2$ by the monic linear polynomial $h(x) = x - 2$.

Solution. Since the divisor is monic, we write it as $1 - 2$ in the table.
Thus the quotient polynomial is the quadratic polynomial $q(x) = 2x^2 + 8x + 17$ and the remainder is the constant polynomial $r(x) = 32$.   //

Notice that at each stage, we multiply the next term of the quotient by 1 and then subtract the product from itself resulting in a zero in that column of the table. Eliminating the successive vertical displacement for each product, one can choose to write all products on a single line. Typically, one omits the coefficient 1 of the x-term in the divisor and changes the sign of the constant term. In this scenario, Table 4.3 becomes:

$$
\begin{array}{r|rrr}
 & 2\ 4 & 1 & -2 \\
2 & 4 & 16 & 34 \\
\hline
 & 2\ 8 & 17 & 32
\end{array}
$$

One thus reads the coefficients of the quotient polynomial from the third row remembering that the final term is the remainder. This process is known as *Synthetic Division* or *Horner's Scheme* but before a formal definition, let us see why the process works.

Proposition 15 (Linear Divisor Case). *If $f(x)$ is the polynomial of degree n given by*

$$f(x) = a_n x^n + a_{n-1} x^{n-1} + a_{n-2} x^{n-2} + \cdots + a_1 x + a_0 \qquad (4.4.2)$$

Table 4.3 Division by a monic linear term as coefficients

$$
\begin{array}{r}
\ \ 2\ +\ 8\ +17\ +32 \\
1-2)\overline{2\ +\ 4\ +\ 1\ -\ 2} \\
\ \ 2\ -\ 4 \\
\hline
\ 8\ +\ 1\ -\ 2 \\
\ 8\ -16 \\
\hline
17\ -\ 2 \\
17\ -34 \\
\hline
32
\end{array}
$$

and

$$g(x) = x - c$$

where $a_0, \ldots, a_n$ and c are real numbers, then

$$f(x) = (b_{n-1}x^{n-1} + b_{n-2}x^{n-2} + \cdots + b_2x^2 + b_1x + b_0)g(x) + r \qquad (4.4.3)$$

where

$$b_{n-1} = a_n \text{ and } b_k = b_{k+1}c + a_{k+1} \text{ if } 1 \le k \le n - 2 \qquad (4.4.4)$$

and

$$r = b_0c + a_0.$$

Proof. Since $g(x) = x - c$, the right hand side of (4.4.3) is equal to

$$b_{n-1}x^n + b_{n-2}x^{n-1} + \cdots + b_2x^3 + b_1x^2 + b_0x$$

$$- cb_{n-1}x^{n-1} - cb_{n-2}x^{n-2} - \cdots - cb_2x^2 - cb_1x - b_0c + r$$

$$= b_{n-1}x^n + (b_{n-2} - cb_{n-1})x^{n-1} + (b_{n-3} - cb_{n-2})x^{n-2} + \cdots +$$

$$(b_2 - cb_3)x^3 + (b_1 - cb_2)x^2 + (b_0 - cb_1)x + (r - b_0c).$$

Comparing this last equation with (4.4.2) term by term, we start with x^n and conclude that $b_{n-1} = a_n$. The next term (the coefficient of x^{n-1}) is $(b_{n-2} - cb_{n-1})$ or $a_{n-1} = b_{n-2} - cb_{n-1}$, which is $a_{k+1} = b_k - cb_{k+1}$ for $k = n - 2$. The other coefficients follow the same rule as given in (4.4.4) by looking at the formula, term by term. ∎

Definition. Given polynomials $f(x) = a_nx^n + a_{n-1}x^{n-1} + a_{n-2}x^{n-2} + \cdots + a_1x + a_0$ and $g(x) = x - c$, the shortcut division of $f(x)$ by $g(x)$ given by the table: is called **Synthetic Division** (also known as **Horner's Scheme**).

The word "sum" in Table 4.4 below means that b_{n-i} is the sum of the two terms above it or:

$$b_{n-i} = b_{n-i+1}c + a_{n-i+1} \text{ and } r = a_0 + b_0c. \qquad (4.4.5)$$

Table 4.4 Synthetic division (Horner's scheme) by a monic linear polynomial

Coefficient		a_n	a_{n-1}	a_{n-2}	$\cdots$	a_1	a_0
Times	c	0	$b_{n-1} \cdot c$	$b_{n-2} \cdot c$	$\cdots$	$b_1 \cdot c$	$b_0 \cdot c$
Sum		$a_n = b_{n-1}$	b_{n-2}	b_{n-3}		b_0	r

Note that, first of all, the above equation is exactly (4.4.4) with $k = n - i$. Secondly, we are used to starting at the beginning (which is where $n = 1$ or 0) of a sequence and working up in these kinds of algorithms. *This method does the opposite*. It starts with the known coefficients of $f(x)$, $a_0, a_1, \cdots, a_n$, and the constant c of $g(x)$ and first finds b_{n-1} (it is a_n). Then, it computes using (4.4.5), the next "lower" b-coefficient, b_{n-1}, as $b_{n-1} = b_n c + a_n$. Having found b_{n-1}, this gives b_{n-2} using (4.4.5) again. This kind of procedure can be called a "descending computation". The term r is the remainder and $q(x) = b_{n-1} x^{n-1} + b_{n-2} x^{n-2} + \cdots + b_2 x^2 + b_1 x + b_0$ is the quotient of $f(x)$ by $x - c$. The "sum" means that, as in (4.4.5),

$$b_{n-i} = b_{n-i+1} c + a_{n-i+1},$$

(as in (4.4.5)) which is just (for each i between 0 and n) the *sum of the two numbers above* b_{n-i}. It is harder to write the formulas than to do the synthetic division!

Example 100. Use Horner's scheme to find the quotient $q(x)$ and the remainder $r(x)$ if $f(x) = 4x^3 + 8x^2 + x - 1$ is divided by $g(x) = x + 2$.

Solution. Since $c = -2$, the table created by Horner's Scheme is below.

$$
\begin{array}{r|rrrr}
     & 4 & 8 & 1 & -1 \\
-2   &   & -8 & 0 & -2 \\
\hline
     & 4 & 0 & 1 & -3
\end{array}
$$

Hence, $q(x) = 4x^2 + 1$, and r is the constant polynomial $r(x) = -3$.   //

Exercise 56. Do each of the divisions below twice, once using long division and the second time with Synthetic Division. You should be able to read the similarities of the two methods in your work.

(a) $5x^5 + 4x^3 + 2x + x + 2$ by $x - 3$
(b) $x^4 - 8x^2 + 16$ by $x + 2$

Synthetic division can also be used if the divisor is $ax - b$, rather than the monic polynomial $x - c$. If the polynomial $f(x)$ is divided by the general linear polynomial $ax - b$ then

$$f(x) = q(x)(ax - b) + r$$

or

$$f(x) = a q(x) \left(x - \frac{b}{a} \right) + r. \tag{4.4.6}$$

Thus if one wishes to apply Synthetic division to divide a polynomial by a general linear polynomial $ax - b$, the operation is handled by first dividing by the monic $x - \frac{b}{a}$ as shown in Example 101 and using Eq. (4.4.6).

Example 101. Use Horner's scheme to find the quotient $q(x)$ and the remainder $r(x)$ if $f(x) = 9x^3 + 4x^2 - 5x - 2$ is divided by $2x - 1$.

Solution. From the discussion of the last paragraph, we first divide $f(x)$ by $x - \frac{1}{2}$

using the table below.

$$
\begin{array}{r|rrr}
 & 9 & 4 & -5 & -2 \\
\frac{1}{2} & & \frac{9}{2} & \frac{17}{4} & -\frac{3}{8} \\
\hline
 & 9 & \frac{17}{2} & -\frac{3}{4} & -\frac{19}{8}
\end{array}
$$

This last table means, of course, that $f(x)$ divided by $x - \frac{1}{2}$ has quotient $9x^2 + \frac{17}{2}x - \frac{3}{4}$ with remainder $-\frac{19}{8}$ or

$$
f(x) = \left(9x^2 + \frac{17}{2}x - \frac{3}{4}\right)\left(x - \frac{1}{2}\right) + \left(-\frac{19}{8}\right).
$$

However, we are looking for the divisor $2x - 1$, not the divisor $x - \frac{1}{2}$. We now just multiply and divide the first term on the left hand side by 2 to obtain the solution, as we did in the general case of Eq. (4.4.6):

$$
f(x) = \frac{1}{2}\left(9x^2 + \frac{17}{2}x - \frac{3}{4}\right)2\left(x - \frac{1}{2}\right) - \frac{19}{8}
$$

$$
= \left(\frac{9}{2}x^2 + \frac{17}{4}x - \frac{3}{8}\right)(2x - 1) - \frac{19}{8}.
$$

Thus, the quotient of $f(x)$ by $2x - 1$ is $q(x) = \frac{9}{2}x^2 + \frac{17}{4}x - \frac{3}{8}$ and the remainder is $-\frac{19}{8}$.   //

Proposition 16 (Synthetic Division for Linear Polynomial). *If $f(x) = a_n x^n + a_{n-1}x^{n-1} + \cdots + a_1 x + a_0$ is divided by the linear polynomial $g(x)$ then*

A. *(Monic Case) If $g(x) = x - c$ where $c \in \mathbb{R}$ then the quotient of $f(x)$ by $g(x)$ is $q(x) = b_{n-1}x^{n-1} + \cdots + b_0$ and the remainder is $r = a_0 + b_0 c$, where $b_0, \cdots, b_{n-1}$ are defined by Eq. (4.4.5).*

B. *(General Case) If $g(x) = ax - b$ for $a, b \in \mathbb{R}$ then the quotient of $f(x)$ by $g(x)$ and the remainder r are obtained by first using Part A to find the quotient, written $q_0(x)$, and the remainder, r_0, of $f(x)$ by $x - \frac{b}{a}$. Then $q(x) = \frac{1}{a}q_0(x)$ and $r = r_0$.*

Exercise 57. Use Synthetic Division to find the quotient and remainder, $q(x)$ and $r(x)$, of the first polynomial divided by the second.

(a) $4x^4 + x^3 + 3x^2 - 4x + 1$, by $4x - 1$
(b) $3x^4 + x^2 - 4x + 5$, by $5x + 1$

Horner's scheme can also be extended to a more general case in which the divisor is a quadratic polynomial. If $f(x) = a_n x^n + a_{n-1}x^{n-1} + \cdots + a_0$ is divided by the monic quadratic $x^2 - bx - c$, we may write a table, similar to Table 4.4 in the linear divisor case, for polynomials of any degree divided by a monic quadratic. Before we give that result (Table 4.5), we will concentrate on the case in which $f(x)$ is a quartic polynomial to simplify the notation. All of the conceptual issues for arbitrary polynomials are already present if $f(x)$ has degree four.

Proposition 17 (Monic Quadratic Divisor Case). *If $f(x)$ is a quartic polynomial given by*

$$f(x) = a_4x^4 + a_3x^3 + a_2x^2 + a_1x + a_0$$

and $g(x)$ is the monic quadratic

$$g(x) = x^2 - bx - c$$

where a_0, a_1, a_2, a_3, a_4, b, $\& c \in \mathbb{R}$ then

$$f(x) = (b_2x^2 + b_1x + b_0)g(x) + r(x) \tag{4.4.7}$$

where $r(x) = r_1x + r_0$ and

$$b_2 = a_4$$
$$b_1 = a_3 + bb_2$$
$$b_0 = a_2 + bb_1 + cb_2 \tag{4.4.8}$$
$$r_1 = a_1 + bb_0 + cb_1$$
$$r_0 = a_0 + cb_0.$$

Thus the quotient of $f(x)$ by $g(x)$ is $q(x) = b_2x^2 + b_1x + b_0$ and the remainder is $r(x) = r_0 + r_1x$.

Proof. Following the same steps as in Proposition 15 (*i.e.* start by expanding the right hand side of (4.4.7)) we obtain,

$$f(x) = b_2x^4 + (b_1 - bb_2)x^3 + (b_0 - bb_1 - cb_2)x^2 + (r_1 - bb_0 - cb_1)x + (r_0 - cb_0). \tag{4.4.9}$$

Equating coefficients of $f(x)$ to those on the right hand of (4.4.9) yields the result, Eq. (4.4.8). ∎

What would be the table similar to Table 4.4 of the linear divisor case for a quartic divided by a monic quadratic? Proposition 17 gives Table 4.5.

Table 4.5 Synthetic division of quartic by monic quadratic

Coefficient		a_4	a_3	a_2	a_1	a_0
Times	b	0	$b \cdot b_2$	$b \cdot b_1$	$b \cdot b_0$	0
Times	c	0	0	$c \cdot b_2$	$c \cdot b_1$	$c \cdot b_0$
Sum		b_2	b_1	b_0	r_1	r_0

Table 4.6 Synthetic division of general polynomial by monic quadratic

Coefficient		a_n	a_{n-1}	a_{n-2}	a_{n-3}	$\cdots$	a_2	a_1	a_0
Times	b	0	$b \cdot b_{n-2}$	$b \cdot b_{n-3}$	$b \cdot b_{n-4}$	$\cdots$	$b \cdot b_1$	$b \cdot b_0$	0
Times	c	0	0	$c \cdot b_{n-2}$	$c \cdot b_{n-3}$	$\cdots$	$c \cdot b_2$	$c \cdot b_1$	$c \cdot b_0$
Sum		$a_n = b_{n-2}$	b_{n-3}	b_{n-4}	b_{n-5}	$\cdots$	b_0	r_1	r_0

Once again, "sum" means that the entry is the sum of the entries above it. Note how each element of the second row gets multiplied by b and those of the third row get multiplied by c. Also notice how the indices of the a_i goes down by one across the top, how that of the b_i goes down across the second row (after being multiplied by b) and that of the b_i goes down by one in the last row (after being multiplied by c). Also, each of the rows "starts" one entry over as we move down rows. Furthermore the symmetry of the table can be thought of as a reason why there is a zero at the end of the second row. After all, there is no b_{-1} (which would be the next coefficient to appear if that makes sense) so putting zero keeps the symmetry. Of course, the last sentence is not a mathematical statement—the mathematics is contained in Proposition 17.

To divide a polynomial of degree n by a monic quadratic, the table will simply need to have more columns to compensate for the additional terms. Once again, $r_1 x + r_0$ is the remainder, and $q(x) = b_{n-2} x^{n-2} + \cdots + b_1 x + b_0$ is the quotient of $f(x) = a_n x^n + \cdots + a_0$ by $g(x) = x^2 - bx - c$ (Table 4.6).

Example 102. Find the quotient $q(x)$ and the remainder $r(x)$ if $f(x) = x^4 - 2x^3 + x^2 - 5x + 4$ is divided by $g(x) = x^2 - x + 1$ by using Horner's scheme.

Solution. Table 4.5 for this case is:

$$
\begin{array}{r|rrr|rr}
 & 1 & -2 & 1 & -5 & 4 \\
1 & 0 & 1 & -1 & -1 & 0 \\
-1 & 0 & 0 & -1 & 1 & 1 \\
\hline
 & 1 & -1 & -1 & -5 & 5
\end{array}
$$

Hence $q(x) = x^2 - x - 1$, $r(x) = -5x + 5$. We can always check with long division (or by multiplying $q(x)g(x)$ and adding $r(x)$).

$$
\begin{array}{r}
x^2 - x - 1 \\
x^2 - x + 1\overline{)x^4 - 2x^3 + x^2 - 5x + 4} \\
\underline{x^4 - x^3 + x^2} \\
-x^3 \quad\quad - 5x \\
\underline{-x^3 + x^2 - x} \\
-x^2 - 4x + 4 \\
\underline{-x^2 + x - 1} \\
- 5x + 5
\end{array}
$$

$$//$$

Exercise 58. Use Horner's scheme (Synthetic Division) to find the quotient and remainder if the first polynomial is divided by the second. Check by multiplying the quotient by the divisor and adding the remainder.

A. $x^4 + 2x^3 - x^2 + 5x - 2$ by $x^2 + 2x - 1$.
B. $x^4 + 4x^3 - 2x^2 - 12x - 3$ by $x^2 - 3$.
C. $3x^4 + 4x^2 - 5$ by $x^2 + x$.

One underlying assumption of Proposition 17 is that the quadratic is a monic polynomial. Similar to the case when dividing by a general linear polynomial, when dividing by a general quadratic polynomial $g(x) = ax^2 + bx + c$, one can factor out the lead coefficient and use as a divisor $x^2 + \frac{b}{a}x + \frac{c}{a}$ instead before dividing the resulting quotient by the coefficient a. The following proposition formalizes this process for division by a non-monic quadratic polynomial.

Proposition 18. *Let $f(x)$ be a polynomial and $g(x) = ax^2 - bx - c$ be a quadratic. Then*

A. *the remainder of $f(x)$ on division by $g(x)$ is the same as the remainder of $f(x)$ on division by $x^2 - \frac{b}{a}x - \frac{c}{a}$*
B. *the quotient of $f(x)$ on division by $g(x)$ is $\frac{1}{a}$ times the quotient of $f(x)$ divided by $x^2 - \frac{b}{a}x - \frac{c}{a}$.*

Proof. Let $q(x)$ be the remainder of $f(x)$ divided by $g(x) = ax^2 - bx - c$ and $r(x) = dx + e$ the remainder. If the polynomial $f(x)$ is divided by the polynomial $ax^2 - bx - c$, then we have

$$
f(x) = q(x)(ax^2 - bx - c) + (dx + e)
$$
$$
= aq(x)(x^2 - \tfrac{b}{a}x - \tfrac{c}{a}) + dx + e. \tag{4.4.10}
$$

Therefore $q(x)$, the quotient of $f(x)$ by $g(x)$, is given by dividing the quotient of $f(x)$ when by $x^2 + \frac{b}{a}x - \frac{c}{a}$ by a; in either case the remainders are the same whether we divide by $g(x)$ or $x^2 - \frac{b}{a}x - \frac{c}{a}$. ∎

Example 103. Using Horner's Method, if $4x^4 + 2x^3 - 4x^2 + 6x - 1$ is divided by $g(x) = 2x^2 - x - 1$, then what are the remainder and quotient? Compare to Exercise 51 of Section 4.2.

Solution. We apply Horner's scheme again with the divisor $h(x) = x^2 - \frac{1}{2}x - \frac{1}{2}$, instead of $g(x)$. Of course, $h(x) = g(x)/2$.

$$
\begin{array}{r|rrrrr}
 & 4 & 2 & -4 & 6 & -1 \\
\frac{1}{2} & & 2 & 2 & 0 & \\
\frac{1}{2} & & & 2 & 2 & 0 \\
\hline
 & 4 & 4 & 0 & 8 & -1
\end{array}
\tag{4.4.11}
$$

The remainder of $4x^4 + 2x^3 - 4x^2 + 6x - 1$ on division by $g(x)$ or $h(x)$ is thus $8x - 1$ from Proposition 18A. Using part B of that same proposition, the quotient on division by $h(x)$ is $\frac{1}{2}$ of $4x^2 + 2x$ or $q(x) = 2x^2 + 2x$, thanks again to Eq. (4.4.10). The problem can be done directly without use of Synthetic Division as Eq. (4.4.12) shows.

$$
\begin{array}{r}
2x^2 + 2x \\
2x^2 - x - 1 \overline{)\, 4x^4 + 2x^3 - 4x^2 + 6x - 1} \\
\underline{-4x^4 + 2x^3 + 2x^2} \\
4x^3 - 2x^2 + 6x \\
\underline{-4x^3 + 2x^2 + 2x} \\
8x - 1
\end{array}
\tag{4.4.12}
$$

Proposition 18A shows that $8x - 1$ is also the remainder of $f(x)$ on division by $g(x)$ (as well as by $h(x)$). //

Exercise 59. In Example 103, why does the table show $4\ 4\ 0\ 8\ -1$ which would seem to imply that the quotient is $4x^2 + 4x$ whereas the computation (4.4.12) shows the quotient to be $2x^2 + 2x$?

The Synthetic Division can be used to write a given polynomial as the polynomial in $ax + b$.

Example 104. Let $f(x) = x^4 - 3x^3 + x^2 + x + 19$. Express $f(x)$ as the polynomial in $x - 2$, i.e. find a, b, c, d, e such that

$$
f(x) = a(x-2)^4 + b(x-2)^3 + c(x-2)^2 + d(x-2) + e.
$$

Solution. Using the Synthetic Division, we have

$$
\begin{array}{r|rrrrr}
 & 1 & -3 & 1 & 1 & 19 \\
2 & & 2 & -2 & -2 & -2 \\
\hline
 & 1 & -1 & -1 & -1 & ① \\
2 & & 2 & 2 & 2 & \\
\hline
 & 1 & 1 & 1 & ① & \\
2 & & 2 & 6 & & \\
\hline
 & 1 & 3 & ⑦ & & \\
2 & & 2 & & & \\
\hline
 & ① & ⑤ & & &
\end{array}
$$

Thus $f(x) = (x-2)^4 + 5(x-2)^3 + 7(x-2)^2 + (x-2) + 17$.

$$\boxed{a = 1, \quad b = 5, \quad c = 7, \quad d = 1, \quad e = 17}$$

Notice that these values are the same coefficients of the Taylor series about 2.

$$f(x) = x^4 - 3x^3 + x^2 + x + 19 \Rightarrow f(2) = 16 - 3(8) + 4 + 2 + 19 = 17$$
$$f'(x) = 4x^3 - 9x^2 + 2x + 1 \Rightarrow f'(2) = 4(8) - 9(4) + 2(2) + 1 = 1$$
$$f''(x) = 12x^2 - 18x + 2 \Rightarrow f''(2) = 12(4) - 18(2) + 2 = 14$$
$$f'''(x) = 24x - 18 \Rightarrow f'''(2) = 24(2) - 18 = 30$$
$$f^{(4)}(x) = 24 \Rightarrow f^{(4)}(2) = 24$$

By Taylor's Theorem, we know that

$$f(x) = \sum_{n=0}^{\infty} \frac{f^{(n)}(2)}{n!}(x-2)^n = \frac{17}{0!} + \frac{1}{1!}(x-2) + \frac{14}{2!}(x-2)^2$$
$$+ \frac{30}{3!}(x-2)^3 + \frac{24}{4!}(x-2)^4.$$
$$= 17 + (x-2) + 7(x-2)^2 + 5(x-2)^3 + (x-2)^4.$$

//

Exercise 60. Let $f(x) = 4x^3 + 5x^2 - 7x + 4$. Express $f(x)$ as a polynomial in $(x-3)$.

Problem Set 4.4:

1. Use only the coefficients in long division to find the quotient $q(x)$ and the remainder $r(x)$ when the first polynomial is divided by the second. Before you start, decide what the degree of the quotient will be.

    (a) $5x^4 - 3x^3 - x^2 + x - 5$ by $2x^2 + x + 1$
    (b) $2x^5 + 3x^4 + x - 5$ by $2x^2 - x + 1$

2. Use Horner's scheme to complete the following polynomial divisions by a monic linear polynomial.

    (a) $3x^3 - 2x^2 - 11x + 12$ by $x - 4$.
    (b) $x^4 + 2x^3 - x^2 + x + 6$ by $x + 2$.
    (c) $x^5 - 3x^4 + x^3 - 5x^2 + x - 7$ by $x - 2$.

3. Use Synthetic division to complete the following polynomial divisions by a general linear polynomial.

    (a) $3x^3 - 2x + x - 1$ by $3x - 1$.
    (b) $2x^5 - 6x^4 + 6x^3 - 4x^2 + 2x - 7$ by $2x + 1$.
    (c) $8x^3 - 24x - 1$ by $4x - 1$.

4. Use Synthetic division to complete the following polynomial divisions by a monic quadratic.

    (a) $x^3 + x^2 - 5x + 3$ by $x^2 - 1$
    (b) $x^4 + 5x^3 + 20x^2 + 80x + 64$ by $x^2 + 5x + 4$
    (c) $x^4 - 8x^3 + 15x^2 + 8x - 16$ by $x^2 + 2x + 1$

5. Use **Horner's Scheme** to find the quotient and remainder of $f(x)$ divided by $g(x)$ if

    (a) $f(x) = x^4 + 4x^3 - 2x^2 - 12x - 3$ by $g(x) = -x^2 - 3$
    (b) $f(x) = 6x^4 + 2x^3 - 3x^2 - x$ by $g(x) = 2x^2 - 1$
    (c) What would happen if you did Part A or B using long division instead of Horner's Scheme. (Hint: This is a trick question!)

6. (Synthetic Division of a cubic by a monic quadratic) Let a_3, a_2, a_1, b and c be real numbers and let $f(x) = a_3 x^3 + a_2 x^2 + a_1 x + a_0$ and $g(x) = x^2 - bx - c$. If $q(x) = b_2 x^2 + b_1 x + b_0$ and $r(x) = r_1 x + r_0$ for fixed real numbers $b_0, b_1,$ b_2, r_0, r_1 are such that $q(x)$ is the quotient of $f(x)$ by $g(x)$ is the remainder. What is the formula for $b_2, b_1, b_0, r_1,$ and r_0 in terms of the other coefficients similar to (4.4.8). (Hint: Mimic Proposition 17 recognizing that in this exercise $f(x)$ is cubic, not quartic.)

7. By using the Synthetic Division, find a, b, c, d and e.

   (a) $4x^3 - 3x + 5 = a(2x - 1)^3 + b(2x - 1)^2 + c(2x - 1) + d$. (Answer: $a = 1/2, b = 3/2, c = 0, d = 4$)
   (b) $x^4 - 3x^3 + x^2 + x + 19 = a(x-2)^4 + b(x-2)^3 + c(x-2)^2 + d(x-2) + e$. (Answer: $a = 1, b = 5, c = 7, d = 1, e = 17$)

8. Do #7 by using Taylor's Theorem

Chapter 5
Factoring Polynomials, Their Roots, and Some Applications

Introduction This chapter focuses on the relationship between polynomials with real coefficients, roots which may be complex numbers or rational roots, and GCD and LCM for polynomials.

5.1 Factoring Polynomials and Their Roots

Overview In Chap. 4, we discussed divisibility of polynomials and the resulting quotient and remainder. Factoring polynomials means looking for quotients where the remainder is zero. Thus, the factoring of polynomials is intimately connected with the roots of polynomials. This relationship is explored extensively in middle and high school algebra to solve problems involving the roots of polynomials, but many students do not realize the conceptual basis for the connection between finding roots and factoring.

In this section, we will see how easy it is to show the connection between factoring and finding the roots of a polynomial and then present a variety of problems involving the relationship between a polynomial and its roots. We will be using Greek letters (α, β, γ, ...) for the roots of polynomials. There is no assumption that these real numbers are integers or rational unless explicitly stated (as in Sect. 5.2).

The Remainder Theorem (Theorem 16 of Sect. 4.3) has an important consequence which is a corollary showing the connection between factoring a polynomial and its roots (or zeros). The corollary is so important that we describe it as the Factor or Root Theorem. That relationship was hinted at in Example 96 of Sect. 4.3.

We will first write the definition of "factor", which mimics the definition for integers. Recall that for integers a and b, if there is an integer c with $b = ca$, then a is called a *factor* of b.

© Springer International Publishing Switzerland 2015
R.S. Millman et al., *Problems and Proofs in Numbers and Algebra*,
DOI 10.1007/978-3-319-14427-6_5

Definition. A polynomial $g(x)$ is a FACTOR of the polynomial $f(x)$ if there is a polynomial $q(x)$ such that $f(x) = q(x)g(x)$. In this case, we also say that $g(x)$ DIVIDES $f(x)$ and write $g(x)|f(x)$.

The definition above is equivalent to saying $g(x)$ is a factor of $f(x)$ exactly when there is a polynomial $q(x)$ such that $f(x) = q(x)g(x)$. In this form the equation for factoring looks very similar to the equation we gain from factoring integers. We say that $g(x)$ DIVIDES $f(x)$ if there is a polynomial $q(x)$ such that $f(x) = q(x)g(x)$.

Theorem 18 (Factor or Root Theorem). *Let $f(x)$ be a polynomial and $c \in \mathbb{R}$. Then $f(x)$ has the linear polynomial $x - c$ as a factor if and only if c is a zero of $f(x)$, that is $f(c) = 0$.*

Proof. By the Remainder Theorem, we have

$$f(x) = q(x)(x - c) + f(c).$$

Hence $(x - c)$ is a factor of $f(x)$ if and only if $f(c) = 0$.   ∎

Example 105. If $f(x) = 3x^3 - 9x^2 + kx - 12$ has a factor $x - 3$, then what is the value of k? Find two ways to solve this example.

Solution.

First Solution. By the Factor Theorem, we must have $f(3) = 0$. Thus $3 \cdot 3^3 - 9 \cdot 3^2 + 3 \cdot k - 12 = 0$ or $81 - 81 + 3k - 12 = 0$. We conclude that $k = 4$.

Second Solution. Another way to solve this problem is to divide $f(x)$ by $x - 3$ to obtain

$$f(x) = (3x^2 + k)(x - 3) + (3k - 12).$$

For $x - 3$ to be a factor of $f(x)$, the remainder of $f(x)$ on division by $(x - 3)$ must be zero. Hence $3k - 12 = 0$, which means that $k = 4$ as in the previous solution.

       //

Example 106. Let $f(x) = 345x^5 - 655x^4 - 50x^2 + 9x + 1$. Use the Root Theorem to determine whether $x + 1$ is a factor of $f(x)$.

Solution. Since

$$f(-1) = 345(-1)^5 - 655(-1)^4 - 50(-1)^2 + 9(-1) + 1 = -1058 \neq 0,$$

it follows from the Root Theorem that $x + 1$ is not a factor of $f(x)$.   //

Exercise 61.

(a) If $f(x) = x^4 - 3ax^2 + bx + 4$ has $(x + 1)$ and $(x - 2)$ as two factors, find a and b.

(b) Knowing the solution to Part(a), are there other linear factors of $f(x)$ with real coefficients besides $(x + 1)$ and $(x - 2)$?

Theorem 19. *If $ax^2 + bx + c$ is a quadratic equation with real coefficients, then it has real roots if and only if $b^2 - 4ac \geq 0$. Furthermore, if α and β are those roots, then*

$$\alpha + \beta = -\frac{b}{a} \text{ and } \alpha\beta = \frac{c}{a} \tag{5.1.1}$$

Proof. For the first sentence of the Theorem, if α and β are the roots then $(x - \alpha)$ and $(x - \beta)$ are factors of $f(x) = ax^2 + bx + c$ and so $(x - \alpha)(x - \beta) = x^2 - (\alpha + \beta)x + \alpha\beta$ also is a factor of $f(x)$. Comparing quadratic terms,

$$a(x - \alpha)(x - \beta) = a(x^2 - (\alpha + \beta)x + \alpha\beta) = ax^2 + bx + c$$

and Eq. (5.1.1) are immediate.

For the second sentence, we will use the standard "complete the square" approach and then derive the quadratic equation as a corollary. The roots α and β of the given equation occur exactly when

$$x^2 + \frac{b}{a}x + \frac{c}{a} = 0. \tag{5.1.2}$$

Equation (5.1.2) has the same solutions by adding and subtracting $\left(\frac{b}{2a}\right)^2$

$$\left(x + \frac{b}{2a}\right)^2 - \frac{b^2}{4a^2} + \frac{c}{a} = 0$$

$$\left(x + \frac{b}{2a}\right)^2 = \frac{b^2}{4a^2} - \frac{c}{a} = \frac{b^2 - 4ac}{4a^2}. \tag{5.1.3}$$

Simplifying the previous equation, this proof shows that Eq. (5.1.3) has a real root if and only if its right hand side is non-negative, which happens exactly when $b^2 - 4ac \geq 0$. ∎

Note: The quadratic $ax^2 + bx + c$ has complex roots if and only if $b^2 - 4ac < 0$.

Implicit in this proof is the quadratic formula from high school which we state separately. The proof follows immediately from Eq. (5.1.3) by taking the square of both sides of the equation.

Corollary 8. *The quadratic equation with real coefficients, a, b and c*

$$ax^2 + bx + c = 0$$

has roots

$$x = \frac{-b \pm \sqrt{b^2 - 4ac}}{2a}.$$

Because of the previous corollary, the Proposition 19 is immediate.

Proposition 19 (Rational Roots for Quadratic Equations). *The quadratic equation*

$$ax^2 + bx + c = 0$$

has rational roots if and only if $b^2 - 4ac$ is a perfect square.

Using our newly developed language from Chap. 5, if $b^2 - 4ac \geq 0$ then the roots α and β are real and

$$\alpha = \frac{-b + \sqrt{b^2 - 4ac}}{2a} \text{ and } \beta = \frac{-b - \sqrt{b^2 - 4ac}}{2a}.$$

The quantity $b^2 - 4ac$ comes up so often in algebra it is given a special name.

Definition. If $f(x) = ax^2 + bx + c$ is a quadratic, then the DISCRIMINANT of f, $\Delta(f)$, is given by

$$\Delta(f) = b^2 - 4ac.$$

It is usual to omit f and write Δ if there is only one polynomial involved.

Once complex numbers are discussed, the results above are interpreted as $\Delta(f) \geq 0$ means that the roots of f are real and $\Delta(f) < 0$ means that they are complex (with a non-zero imaginary part).

Example 107. Let α, β be the roots of $x^2 + 4x - 9 = 0$. Find the values of $\alpha^2\beta + \alpha\beta^2$ and $\alpha^3 + \beta^3$ without finding α and β.

Solution. Because of Eq. (5.1.1), $\alpha + \beta = -4$ and $\alpha\beta = -9$. Thus,

$$\alpha^2\beta + \alpha\beta^2 = \alpha\beta(\alpha + \beta) = (-9)(-4) = 36.$$

Similarly,

$$\alpha^3 + \beta^3 = (\alpha + \beta)^3 - 3\alpha\beta(\alpha + \beta) = (-4)^3 - 3(-9)(-4) = -172. \quad \text{//}$$

Exercise 62. Let α, β be the roots of $x^2 + 6x - 120 = 0$. Find the values of $\frac{\beta}{\alpha} + \frac{\alpha}{\beta}$ and $\alpha^2 + \beta^2$ without finding α and β.

Example 108. Let α, β be the roots of $3x^2 + 6x + 9 = 0$. Find a quadratic equation whose two roots are $\frac{\beta}{\alpha}$ and $\frac{\alpha}{\beta}$.

Solution. Since α and β are the roots of $3x^2 + 6x + 9 = 0$, we have $\alpha + \beta = -2$ and $\alpha\beta = 3$. The quadratic polynomial we are looking for is $x^2 + bx + c$ where $-b = \frac{\beta}{\alpha} + \frac{\alpha}{\beta}$ and $c = \left(\frac{\beta}{\alpha}\right)\left(\frac{\alpha}{\beta}\right) = 1$ [see Eq. (5.1.1)]. Now

$$-b = \frac{\beta}{\alpha} + \frac{\alpha}{\beta} = \frac{\beta^2 + \alpha^2}{\alpha\beta} = \frac{(\alpha + \beta)^2 - 2\alpha\beta}{\alpha\beta} = \frac{(-2)^2 - 2 \cdot 3}{3} = \frac{-2}{3},$$

so the monic quadratic equation satisfying the conditions of the problem is therefore $x^2 - \frac{2}{3}x + 1 = 0$. An alternative solution is $3x^2 - 2x + 3 = 0$. //

Exercise 63. Let α, β be the roots of $2x^2 + 3x + 6 = 0$. Find a quadratic equation whose roots are 2α and 2β.

Exercise 64. If $w \in \mathbb{R}$ and the equation

$$x^2 + wx + 5 = 0$$

has two roots α and β, find the values of the quantities below without first calculating α and β.

(a) $\frac{\alpha}{\beta} + \frac{\beta}{\alpha}$ (b) $\alpha^2 + \beta^2$ (c) $\alpha^3 + \beta^3$ (d) $\frac{1+\alpha^{20}}{1+\beta^{20}}$

Proposition 20. *The quadratic equation $ax^2 + bx + c = 0$ has two equal roots if and only if $b^2 = 4ac$ (that is, the discriminant of the quadratic is zero).*

Proof. Let α, β be the roots. Since $\alpha + \beta = -\frac{b}{a}$ and $\alpha\beta = \frac{c}{a}$, we have

$$b^2 - 4ac = a^2(\alpha + \beta)^2 - 4a(\alpha + \beta)$$
$$= a^2[\alpha^2 - 2\alpha\beta + \beta^2] = a^2(\alpha - \beta)^2$$

Thus $b^2 - 4ac = 0$ if and only if $\alpha = \beta$. ∎

This Proposition can also be seen immediately through the quadratic formula, since the roots are

$$x = \frac{-b \pm \sqrt{b^2 - 4ac}}{2a}.$$

If $b^2 - 4ac$ is not zero then there are either two different roots ($b^2 - 4ac > 0$) or no real roots ($b^2 - 4ac < 0$).

Example 109. Suppose $(m + 2)x^2 + 4x + (m - 1) = 0$ has two distinct real roots and $m \in \mathbb{Z}$. Find all possible values of m.

Solution. Since $(m + 2)x^2 + 4x + (m - 1) = 0$ has two distinct real roots, we know that $\Delta = \Delta(m)$ (which is a function of the integer m) must be positive. We can write this condition as

$$\Delta(m) = 16 - 4(m + 2)(m - 1) > 0 \text{ or}$$

$$\Delta = 16 - 4(m + 2)(m - 1) = -4(m^2 + m - 6) = -4(m + 3)(m - 2) > 0.$$

Thus $(m + 3)(m - 2) < 0$, which means that the two factors must have different signs and so $-3 < m < 2$. Note that $m = -2$ is not a possible value for m since $(m + 2)x^2 + 4x + (m - 1) = 0$ would then be a linear equation having only one root $x = \frac{3}{4}$. Because the problem states that m must be an integer, the only possibilities are $m = -1, 0$ or 1. It is easy to check that each of these values of m works.   //

Example 110. Let a, b, and c be the lengths of the three sides of the triangle $\triangle ABC$. If $a \neq c$, show that the equation $(a - c)x^2 + 2bx - (c - a) = 0$ has two distinct real roots.

Solution. Our plan is to look at the discriminant of the quadratic given in the Example and show it is positive. By definition,

$$\Delta = 4b^2 + 4(a - c)(c - a)$$

$$= 4(b^2 - (a - c)^2) = 4(b - (a - c))(b + (a - c)).$$

Thus

$$\Delta = 4((b + c) - a)((b + a) - c). \tag{5.1.4}$$

Since the sum of the lengths of any two sides of a triangle is larger than the third, both factors of Eq. (5.1.4) are positive so that $\Delta > 0$ and Theorem 19 of this section shows that the equation has distinct real roots.   //

Exercise 65. Assume that $(a + b)x^2 + 2cx + (a - b) = 0$ has two equal roots. Prove that $\triangle ABC$ is a right triangle (where a, b, and c are the lengths of the sides of $\triangle ABC$).

Problem Set 5.1:

Note: You could use Synthetic Division or direct division by polynomials if applicable.

1. Is $(x - 2)$ a factor of $x^3 - 4x^2 + 5x + 1$?
2. Is $(x + 1)$ a factor of $x^3 - 5x^2 + 2x + 8$?
3. Is $(x - 3)$ a factor of $x^3 - 3x^2 - 13x + 15$?
4. Is $(x - 1)$ a factor of $2x^3 - 3x^2 + 14x - 1$?
5. Is $(2x - 3)$ a factor of $6x^3 - 19x^2 - 7x - 12$?
6. Find s so that $x + 2$ is a factor of $x^4 - 3x^3 + sx^2 + x + 5$.

7. Find t so that $x + 3$ is a factor of $x^4 - 7x^3 + tx^2 - 18x - 9$.

8. Find r so that $x + 5$ is a factor of $x^4 + 3x^3 + rx^2 - 33x + 10$.

9. Find k such that $x + 1$ is a factor of $2x^3 + 3x^2 - kx + 10$.

10. Find r and s so that $f(x) = x^3 + rx^2 + sx + 7$ has $x - 3$ and $x + 4$ as factors.

11. Find k and t so that $f(x) = x^3 + kx^2 + tx + 16$ has $x + 2$ and $x - 4$ as factors.

12. Find m and n so that $f(x) = x^3 + mx^2 + nx - 60$ has $x + 5$ and $x + 3$ as factors.

13. An eighth-grade student, Bluma, is trying to solve the equation $5x^2 - 27x + 10 = 0$. She first rewrites (incorrectly) the equation as

$$x^2 - 27x + 5(10) = 0 \qquad (5.1.5)$$

and factors that equation to get $x = 2$ or 25. Then she realizes that she multiplied the constant term by five and so she divides her answers to Eq. (5.1.5) by 5 and claims the solutions of the original are $\frac{2}{5}$ and 5. Are these solutions correct? Whether her method is correct or not is explored in Problem 14.

14. (continuation of Problem 13) Let $f(x) = ax^2 + bx + c$ and let $g(x) = x^2 + bx + ac$. If α and β are the roots of $g(x)$, prove that the roots of $f(x)$ are $\frac{\alpha}{a}$ and $\frac{\beta}{a}$. (This means that Bluma's procedure in Problem 13 is correct. See Clark and Millman [2].)

15. Let $(x + a)(x + b) + (x + b)(x + c) + (x + c)(x + a) = 0$ have two equal roots and assume that a, b, and c are the lengths of the sides of $\triangle ABC$. Then what kind of triangle is $\triangle ABC$. (Prove your assertion.)

16. Let $f(x)$ be the monic cubic polynomial $f(x) = x^3 + ax^2 + bx + c$. If $f(x)$ has roots α, β, and γ, find formulas for a, b, and c in terms of α, β, and γ.

17. If the cubic $f(x)$ of Problem 16 is not monic, how would you modify your formulas?

18. If $m, k \in \mathbb{Z}$ and the quadratic equation

$$x^2 - 4(m - 1)x + (3m^2 - 2m + 4k) = 0$$

has two equal solutions, what are the possible values for m and k?

19. If $k, m \in \mathbb{Z}$, for what values of m will the equation

$$x^2 - 4(m - 1)x + (3m^2 - 2m + 4k) = 0$$

have two equal solutions?

5.2 Rational Roots of Polynomials

Overview The Factor or Root Theorem of the previous section (Theorem 18) provides the condition for existence of a root of a polynomial $f(x)$. However, it does not give a way to *find* such a root. In mathematical language, the Factor Theorem does not give a "constructive condition", meaning that there is no procedure (or algorithm) contained in the proof of the Factor Theorem to show how to construct a root c such that $x - c$ is a factor of $f(x)$. You can already easily obtain the roots of linear and quadratic equations polynomials (or show that there are no real ones).

Rather than talking about a test for roots, this section is dedicated to presenting a constructive method for actually *finding rational roots of polynomials whose coefficients are all integer-valued* (the Rational Root Test, Theorem 20 below). The procedure works well when the roots are rational numbers, but gives no information about the roots that are not rational numbers.

Definition. If $f(x)$ is a polynomial and r is a rational number with $f(r) = 0$, then r is called a RATIONAL ROOT of $f(x)$.

Notation. *We denote by $\mathbb{Z}[x]$, the set of all polynomials in x, each of whose coefficients are integers. We denote by $\mathbb{Q}[x]$, the set of all polynomials in x all of whose coefficients are rational numbers. We write $\mathbb{R}[x]$ as the set of all polynomials whose coefficients are real numbers.*

Exercise 66.

(a) Show that if $f(x)$ and $g(x)$ are in $\mathbb{Z}[x]$, then both $f(x) + g(x) \in \mathbb{Z}[x]$ and $f(x)g(x) \in \mathbb{Z}[x]$.
(b) Prove the same statement for $\mathbb{Q}[x]$.

Exercise 67 (Prove or Disprove). If all coefficients of $f(x)$ are integers, then all of the real roots of $f(x)$ must be integers.

The Rational Root Theorem is a simple criterion for a polynomial in $\mathbb{Z}[x]$ to have (or not to have) a rational root. This criterion translates into a five step procedure to find the rational roots of a polynomial with integer coefficients.

Theorem 20 (Rational Root Test). *Let $f(x) = a_0 + a_1 x + \cdots + a_n x^n \in \mathbb{Z}[x]$ be of degree n. If r is the rational number $\frac{s}{t}$, expressed in lowest terms, and r is a rational root of $f(x)$, then $s \mid a_0$ and $t \mid a_n$.*

Proof. Since $\frac{s}{t}$ is a root of $f(x)$ we have

$$f\left(\frac{s}{t}\right) = a_n \left(\frac{s}{t}\right)^n + a_{n-1} \left(\frac{s}{t}\right)^{n-1} + \cdots + a_1 \left(\frac{s}{t}\right) + a_0 = 0. \qquad (5.2.1)$$

This implies that

$$a_0 t^n + a_1 s t^{n-1} + \cdots + a_{n-1} s^{n-1} t + a_n s^n = 0.$$

Therefore

$$t(a_0 t^{n-1} + a_1 s t^{n-2} + \cdots + a_n s^{n-1}) = -a_n s^n. \qquad (5.2.2)$$

This shows that $t | a_n s^n$. But because s and t are expressed in lowest terms, $\mathrm{GCD}(s, t) = 1$ [see Eq. (2.1.4)]. By Proposition 6, we obtain $t | a_n$ from Eq. (5.2.2) since t must divide either a_n or s. Similarly we can use, from Eq. (5.2.1),

$$a_1 s t^{n-1} + \cdots + a_{n-1} s^{n-1} t + a_n s^n = -a_0 t^n.$$

The analogous argument shows that $s | a_0$.   ∎

Remark. Stated simply, if the rational number in lowest terms is a root of the polynomial, its numerator divides the constant term and its denominator divides the coefficient of the degree of the polynomial.

Exercise 68. Prove the last assertion of Theorem 20.

It is important to understand that the Rational Root Test gives only a necessary condition for $f(x) \in \mathbb{Z}[x]$ to have $r = \frac{s}{t}$ as a root. This means that if r is a root of $f(x)$ then $s | a_0$ and $t | a_n$. It does NOT say that every rational number $\frac{s}{t}$ for which these conditions hold is a root. As an example, if $f(x) = 3x^2 + x + 1$ has $r = \frac{s}{t} = \frac{1}{3}$ as a test, we have $s | a_0$ and $t | a_2$, but $f(\frac{1}{3}) \neq 0$. It is a similar situation to that of the elections in the United States in the sense that the Rational Root Test is like a preliminary election (the "primaries"). Once the primary is over, there is a limited number of candidates (rational numbers in our case) to consider to be elected (to actually be roots).

The advantage of such a necessary condition as the Rational Root Test is that there are not many cases to test for the rational roots. More precisely, we look towards those of the form $r = \frac{s}{t}$ where $s | a_0$ and $t | a_n$. That statement shows the tremendous advantage of this approach because we are then left to check mechanically a few cases as we will illustrate in Example 111.

The Rational Root Test Theorem provides us with the following procedure to find a rational root of $f(x) = a_0 + a_1 x + \cdots + a_n x^n \in \mathbb{Z}[x]$. We know that for a rational root $\frac{s}{t}$, s must divide the constant term, t the leading coefficient, and s and t must be relatively prime.

Rational Root Procedure

Step 1. Find all possible (integer) factors s of the constant term a_0.

Step 2. Find all possible (integer) factors t of the leading coefficient, a_n.

Step 3. Use Steps 1 and 2 to find all possible reduced values of $\frac{s}{t}$.

Step 4. Pick one of the possible roots $r = \frac{s}{t}$ from Step 3 and see whether it is a root (see the remark below). If it is, proceed to Step 5. If not, try another possible rational root from Step 3 until you've successfully found a rational root.

Step 5. To find all rational roots you can either check all the possibilities of Step 3 or use one successful $r = \frac{s}{t}$ to factor $f(x)$. Usually, it is faster to do the factoring and find $g(x)$ where

$$f(x) = (x - r)g(x). \tag{5.2.3}$$

The other rational roots of $f(x)$ are then the rational roots of $g(x)$ of Eq. (5.2.3) and so the same procedure may be applied to $g(x)$. It is also useful to look at the factored polynomial (i.e. at $g(x)$) to see if you can find some other rational roots by factoring or using the quadratic formula. In addition, you may also be able to show there are no further roots by using another technique (e.g. the quadratic formula).

Remark. Step 4 asks you to check whether a possible root $x = r = \frac{s}{t}$ is actually a root or not of $f(x)$. There are two ways to see if r is a root. One is to compute $f(r)$ and check if it is zero and the other is to divide $f(x)$ by $x - r$ using long division or synthetic division and see if the remainder of $f(x)$ on division by $x - r$ is zero or not. The choice is yours—whichever you feel will be faster.

Example 111. Use the Rational Root Test to find rational roots of the equation $15x^2 + 43x + 8 = 0$.

Solution. If $r = \frac{s}{t}$ is a rational solution then $s|8$ and $t|15$. Thus the possibilities are:

$$s = \pm 1, \pm 2, \pm 4, \pm 8 \text{ and } t = \pm 1, \pm 3, \pm 5, \pm 15.$$

The possibilities for r to be tried out are

$$\pm 1, \pm\frac{1}{3}, \pm\frac{1}{5}, \pm\frac{1}{15}, \pm 2, \pm\frac{2}{3}, \pm\frac{2}{5}, \pm\frac{2}{15}, \pm 4, \pm\frac{4}{3}, \pm\frac{4}{5}, \pm\frac{4}{15}, \pm 8, \pm\frac{8}{3}, \pm\frac{8}{5}, \pm\frac{8}{15}$$

which is a rather long list (but is finite). (The answer turns out to be $-\frac{1}{5}$ and $-\frac{8}{3}$.) //

To see how to do this example by factoring (and to review factoring), see the appendix to this section.

Remark. It is important to note that the Rational Root Test is only valid for finding roots that are *rational* numbers. It says nothing about other possibilities. As an example, using the rational root test on $x^2 - 2$ yields the fact that there are no rational roots (the only possibilities are ± 2 and ± 1). Of course, there are two roots $(\pm\sqrt{2})$—they are just not rational numbers. In fact, we have just used the rational root test to reprove the following theorem.

Theorem 21. *The number $\sqrt{2}$ is not a rational number.*

Proof. The discussion of the preceding paragraph shows the only possible rational roots of $x^2 - 2$ are ± 2 or ± 1. As we have assumed that the square root of any non-negative real number is a real number, we may actually assert from Theorem 21 that the $\sqrt{2}$ is irrational. ∎

Remark. In this section we *use the foundation we've built* (the machinery of greatest common divisors from Sect. 2.1) *to solve problems* as we did in Theorem 4 of the first section of Chap. 2.

Remark. We have given the conceptual framework for understanding why the usual approach to solving polynomials in the schools "through factoring of polynomials with integer coefficients" is actually valid.

For Exercises 69 and 70, you should use the Rational Root Theorem in a similar way.

Exercise 69. Prove that $\sqrt{3}$ is irrational.

Exercise 70. If p is a prime number, prove that $\sqrt{p}$ is irrational.

Example 112. Find the rational roots of $f(x) = 2x^3 - 5x^2 - x + 6$.

Solution. We follow the Rational Root Procedure previously outlined, looking for rational roots of the form $\frac{s}{t}$, where s and t are relatively prime.

Step 1. In order to find possible values of s we must find all possible factors of $a_0 = 6$. This results in $s \in \{\pm 1, \pm 2, \pm 3, \pm 6\}$.

Step 2. Find all possible values t we look for factors of $a_n = 2$, which gives $t \in \{\pm 1, \pm 2\}$.

Step 3. To find all possible values of $\frac{s}{t}$. There are only 12 possible rational roots. They are $\pm 1, \pm 2, \pm 3, \pm 6, \pm \frac{1}{2}$, and $\pm \frac{3}{2}$.

Step 4. Choose one of the possible values of $\frac{s}{t}$ from Step 3. We'll try $x = 1$ since it makes the arithmetic easy. Unfortunately for us, $f(1) = 2$ so $x = 1$ is not a root.

Choosing another possible value, we will now try $x = -1$. In this case, $f(-1) = 0$ so -1 is a rational root. We may therefore divide $f(x)$ by $x - (-1) = x + 1$ and obtain

$$f(x) = 2x^3 - 5x^2 - x + 6 = (x + 1)(2x^2 - 7x + 6) + 0$$
$$= (x + 1)(2x - 3)(x - 2)$$

where the quadratic term factors easily as $(2x - 3)(x - 2)$. Therefore the three rational roots of $f(x)$ are $x = -1, \frac{3}{2}$, and 2 since $f(x)$ is of degree 3 and so there are at most 3 roots, we have found all (three) of the roots.   //

It is certainly true that in Example 112, we were "lucky" when our second guess $(x = -1)$ turned out to be a root (and hence we could factor $x + 1$ out from $f(x)$ and finish the problem easily). We'll now revisit the same problem to give a useful and more methodical way to eliminate some of the candidates for s and t. That approach is based on the Factor/Root Theorem (Corollary 18).

Example 113 (Example 112 Revisited). Find the rational roots of $f(x) = 2x^3 - 5x^2 - x + 6$.

Solution. Proceeding to Step 3 above gives 12 possible values of $\frac{s}{t}$. However, from the Factor Theorem we know that $f(x)$ has the linear polynomial $(x - c)$ as a factor if and only if $f(c) = 0$. Thus, if $\frac{s}{t}$ is a root of $f(x)$, there is a polynomial $g(x)$ such that $\left(x - \frac{s}{t}\right) g(x) = f(x)$, or

$$(tx - s)g(x) = tf(x).$$

We therefore see that if $\frac{s}{t}$ is a root of $f(x)$ then, $f(x)$ is divisible by

$$(tx - s). \tag{5.2.4}$$

It is this last statement that allows us a further restriction of the possibilities for $\frac{s}{t}$ in our example. Since $f(1) = 2 - 5 - 1 + 6 = 2$ and $tx - s = t - s$ when $x = 1$, it must be that $(t - s)$ divides 2. That equation allows all of the possibilities for $\frac{s}{t}$ except ± 6 and $-\frac{1}{2}$. We have eliminated 3 of the 12 possibilities. What information does an application of (5.2.4) with $x = -1$ yield?   //

Example 114. Let $a, b \in \mathbb{Z}$. If $f(x) = x^3 + ax^2 + bx + 3$ has 3 distinct rational roots, find the values of a and b.

Solution. By the rational roots theorem, we know that the possible candidates for rational roots are $\pm 1, \pm 3$. Since there are 3 distinct rational roots, and four choices, we write 4 equations with a and b for each of the possibilities $\pm 1, \pm 3$. If $x = 1$ is a root of $f(x) = 0$, then

$$1 + a + b + 3 = 0 \text{ or } a + b = -4. \tag{5.2.5}$$

If $x = -1$ is a root of $f(x) = 0$, then

$$-1 + a - b + 3 = 0 \text{ or } a - b = -2. \tag{5.2.6}$$

If $x = 3$ is a root of $f(x) = 0$, then

$$27 + 9a + 3b + 3 = 0 \text{ or } 3a + b = -10 \tag{5.2.7}$$

If $x = -3$ is a root of $f(x) = 0$, then

$$3a - b = 8. \tag{5.2.8}$$

The first two of these equations gives that $a = -3$ and $b = -1$. Does $a = -3$ and $b = -1$ solve either (5.2.7) (in which case $x = 3$ is a root) or (5.2.8)? Yes, $a = -3$ and $b = -1$ satisfy (5.2.7) so the roots are $x = 1, -1$ and $+3$ and $a = -3$ and $b = -1$.

If we had not been successful in finding the roots in our first guess, $\{1, -1, 3\}$, we would have tried all 4 possibilities until one actually worked.   //

Example 115. Let $m, k \in \mathbb{Z}$. Assume the two real roots of the following equation are equal and in fact are rational numbers.

$$f(x) = mx^2 + (m + 1)x + (k + 1) = 0 \tag{5.2.9}$$

(a) Prove that $k(k + 1)$ must be the square of an integer.
(b) Find the possible values of m and k.

Solution. For Eq. (5.2.9) to have equal roots, its discriminant must be zero. In equation form, this means:

$$\Delta(m) = (m + 1)^2 - 4m(k + 1) = m^2 - (2 + 4k)m + 1 = g(m). \qquad (5.2.10)$$

Let us regard $\Delta(m)$ as a quadratic polynomial in m which we'll call $g(m)$. Since we are searching for solutions to (5.2.10) with m and k integers, it must be that the solutions to $g(m) = 0$ are integers. Thus the quadratic formula applied to (5.2.10) yields

$$m = \frac{(2 + 4k) \pm \sqrt{(2 + 4k)^2 - 4}}{2} \qquad (5.2.11)$$

$$= \frac{(2 + 4k) \pm \sqrt{16(k + k^2)}}{2}$$

$$= (1 + 2k) \pm 2\sqrt{k + k^2}.$$

Thus $k(k + 1)$ must be a perfect square. If $k(k + 1) = 0$, then $m = 1 + 2k$ and either $k = 0$ or $k = -1$. Thus $m = 1$ or $m = -1$ and there are two possibilities:

$$f(x) = x^2 + 2x + 1 \ (k = 0) \text{ or } f(x) = x^2 \ (k = -1). \qquad (5.2.12)$$

We'll now solve Part(b) and show that there are no other solutions. If $k(k + 1)$ is a positive perfect square it has a decomposition into prime factors as

$$k(k + 1) = p_1^{n_1} \cdots p_\ell^{n_\ell} \qquad (5.2.13)$$

where each of the p_i are primes and $n_1, \ldots, n_\ell$ are even integers. Since k and $k + 1$ are relatively prime, they can have no common factors, in particular, no common prime factors. Thus Eq. (5.2.13) shows that both k and $k + 1$ would be the product of primes, each an even number of times. However, both k and $k + 1$ must both be squares (and positive) and this isn't possible. This can happen if they are 1 and 2 (as there are no other consecutive squares). However, when $k = 1, k(k+1) = 2$, which is not a perfect square. Thus Eq. (5.2.13) gives the only solutions to the problem: $k = 0$ and $m = 1$ or $k = -1$ and $m = -1$. $\qquad //$

Once again we have solved a problem by using a foundation built in earlier sections.

Exercise 71. Assume $g(x) = x^3 + ax^2 + bx + 6 \in \mathbb{Z}[x]$ has three distinct rational roots. Find the values of a and b (if any exist).

Problem Set 5.2:

1. Which of the 12 possibilities of Step 3 of Example 112 are eliminated if one uses (5.2.4) with $x = 2$?
2. Find all rational roots of $f(x) = x^3 - 6x^2 - 7x + 60$.
3. Find all rational roots of $h(x) = x^3 + 27$.
4. Find all rational roots of $f(x) = \frac{x^3}{3} - 2x^2 - \frac{7x}{3} + 20$.
5. What are all rational roots of $f(x) = 3x^3 + 5x^2 - 16x - 28$?
6. Show that the polynomial $3x^3 - 3x + 1 = 0$ has no rational roots.
7. Let $f(x) = x^4 - 3x^2 - 6$.

   (a) Find all rational roots of $f(x)$.
   (b) Find all roots of $f(x)$. (Of course, depending on your response to Part A, you may have to use methods that are not in this section.)

8. Using the Rational Root Test, find all *possible* rational roots of the following (you are not expected to find all rational roots, just the possibilities given by the Rational Root Test).

   (a) $9x^4 - 24x^3 - 2x^2 - 24x + 9 = 0$
   (b) $6x^4 + x^3 + 14x^2 - 14x + 3 = 0$
   (c) $2x^3 + 3x^2 - 2x + 3 = 0$

9. Assume $f(x) = 9x^4 + ax^3 + bx^2 + cx + 1 \in \mathbb{Z}[x]$ has 4 distinct rational roots. Find the values of a, b, and c.

5.2.1   Appendix to Sect. 5.2: A Brief Review of Factoring Quadratics

Remark. With some practice you will begin to learn how to pick the right numbers without exhausting all possible cases, thus saving a substantial amount of work.

Remark. The quadratic equations used in this method were carefully selected so that they could be solved by factoring. There are, however, many quadratics that can not be solved in this way. For example, the equation $x^2 - 8x - 3 = 0$.

Example 116. Solve $15x^2 + 43x + 8 = 0$. Here $a = 15, b = 43$, and $c = 8$.

Solution. Steps:

1. Find all possible pairs of factors of a and c. The possibilities are :

$$a = 15 = 15 \cdot 1 = 5 \cdot 3 \text{ and } c = 8 = 8 \cdot 1 = 4 \cdot 2.$$

2. Use of a pair of integers which factors $a = lm$ and write $(\pm lx + \_\,)(\pm mx + \_\,)$. For instance, using the pair of 15 and 1, write the form $(15x + \_\,)(x + \_\,)$.

3. Try all possible pairs of factors of c to fill the blanks in step 2. First we try 8 and 1 in the blanks, *both ways*, namely $(15x+8)(x+1)$ and $(15x+1)(x+8)$. Try 4 and 2 in the blanks, *both ways*, namely $(15x+4)(x+2)$ and $(15x+2)(x+4)$. (You should also do the same for the case in which c is factored as $(-2)(-4)$ or $(-1)(-8)$.)

4. Check the middle term of each product in step 3. If the value of b ($b = 43$ in our example) which is the middle term of the product in not equal to $43x$ for any possibility, then go to step 2 again. It is, unfortunately true that in this example the middle term does not work out for any of these four possibilities.

   We now go back to step two, factorizing $a = 15$ as $15 = 5 \cdot 3$. We consider the pairs 5 and 3, write $(5x+\_)(3x+\_)$. Once again, try 4 and 2 in both ways, and 8 with 1 both ways. We try $(5x+4)(5x+2)$, $(5x+2)(3x+4)$, $(5x+1)(3x+8)$, $(5x+8)(3x+1)$. The middle term of the product of $(5x+1)(3x+8)$ is $43x$, hence $(5x+1)$ and $(3x+8)$ are the solutions for the factors, and we are finished without having to check the factors $(-4)(-2)$ or $(-8)(-1)$.

5. Solve each quadratic equation by factoring:

   (a) $x^2 - 7x + 12 = 0$
   (b) $6x^2 + x - 15 = 0$
   (c) $2x^2 - 7x - 30 = 0$ //

5.3 Greatest Common Divisors and Least Common Multiples for Polynomials

Overview We continue with results for polynomials which are similar to those of Chap. 2 for natural numbers and integers. We first define the GCD and LCM of polynomials which are parallel to that of the GCD and LCM of integers (Table 5.1).

Corresponding to the Division Algorithm for integers (page 17 of Sect. 1.1), there is a division algorithm for polynomials (Theorem 15). Therefore, it should not be surprising that there is also a Euclidean Algorithm for Polynomials to find the greatest common divisor of two polynomials which we will now define and then prove the algorithm. This procedure comes from the division algorithm for polynomials, just as the Euclidean Algorithm for integers (Theorem 7) comes from the Division Algorithm for integers. The proof of Theorem 22 is exactly the same as that of Theorem 7 but with polynomials instead of integers and we will omit it.

Table 5.1 Example 116

15,1	8,1
5,3	4,2
−15,−1	−8,−1
−5,−3	−4,−2

Definition. Let $f(x)$ and $g(x)$ be non-zero polynomials over $\mathbb{R}$. A monic polynomial $d(x)$ over $\mathbb{R}$ is the GREATEST COMMON DIVISOR (GCD) OF $f(x)$ AND $g(x)$ if

(i) $d(x)|f(x)$ and $d(x)|g(x)$
(ii) if $k(x) \in \mathbb{R}[x]$, $k(x)|f(x)$, and $k(x)|g(x)$, then $k(x)|d(x)$.

The greatest common divisor of $f(x)$ and $g(x)$ is usually denoted by $\mathrm{GCD}(f(x), g(x)) = d(x)$ which means that $d(x)$ is a common divisor of both $f(x)$ and $g(x)$ [Part (i)] and $d(x)$ is the largest common divisor [Part (ii)].

The Euclidean Algorithm for polynomials (Theorem 22 below) is similar to the Euclidean Algorithm for finding the GCD of non-zero integers just as the definition of $\mathrm{GCD}(f(x), g(x))$ mimics the definition of the $\mathrm{GCD}(a, b)$ if a and b are non-zero integers.

Definition. If $\mathrm{GCD}(f(x), g(x)) = 1$, then $f(x)$ and $g(x)$ are called RELATIVELY PRIME POLYNOMIALS.

Theorem 22 (Euclidean Algorithm for Polynomials). *Let $f(x)$ and $g(x)$ be polynomials over the $\mathbb{R}$ such that $\deg f(x) \geq \deg g(x)$. Then applying the division algorithm gives a sequence of quotients $(q_0(x)\ q_1(x), \ldots)$ and remainders $(r_0(x)\ r_1(x), \ldots)$ such that*

$$f(x) = q_0(x)g(x) + r_1(x), \qquad \deg r_1(x) < \deg g(x)$$
$$g(x) = q_1(x)r_1(x) + r_2(x), \qquad \deg r_2(x) < \deg r_1(x)$$
$$r_1(x) = q_2(x)r_2(x) + r_3(x), \qquad \deg r_3(x) < \deg r_2(x)$$

$$\cdots$$

$$r_{n-2}(x) = q_{n-1}(x)r_{n-1}(x) + r_n(x), \qquad \deg r_n(x) < \deg r_{n-1}(x)$$
$$r_{n-1}(x) = q_n(x)r_n(x),$$

where $r_n(x)$ is the last non-zero remainder. If a is the leading coefficient of $r_n(x)$ then $\mathrm{GCD}(f(x), g(x)) = a^{-1}r_n(x)$.

Example 117. Find $\mathrm{GCD}(x^5 + 6x^2 - 49x + 42, x^4 + x^3 - 9x^2 - 3x + 18)$.

Solution. We use division of polynomials and follow the procedure and notation of Theorem 22 with $f(x) = x^5 + 6x^2 - 49x + 42$ and $g(x) = x^4 + x^3 - 9x^2 - 3x + 18$. $f(x) = q(x)g(x) + r_1(x)$ is found by division to obtain:

$$x^5 + 6x^2 - 49x + 42 = \underbrace{(x - 1)}_{q_0(x)}(x^4 + x^3 - 9x^2 - 3x + 18) + \underbrace{10x^3 - 70x + 60}_{r_1(x)}$$

Using the notation of the theorem, the above equation gives Theorem 22 for which $g(x)$ divided by $r_1(x)$ and comes up with

Table 5.2 GCD($x^5 + 6x^2 - 49x + 42, x^4 + x^3 - 9x^2 - 3x + 18$)

x	$x^5 + 0x^4 + 0x^3 + 6x^2 - 49x + 42$	$x^4 + x^3 - 9x^2 - 3x + 18$	$\frac{1}{10}x$
	$x^5 + x^4 - 9x^3 - 3x^2 + 18x$	$x^4 + 0x^3 - 7x^2 + 6x$	
-1	$-x^4 + 9x^3 + 9x^2 - 67x + 42$	$x^3 - 2x^2 - 9x + 18$	$\frac{1}{10}$
	$-x^4 - 1x^3 + 9x^2 + 3x - 18$	$x^3 \qquad - 7x + 6$	
$-5x$	$10x^3 + 0x^2 - 70x + 60$	$-2x^2 - 2x + 12$	
	$10x^3 + 10x^2 - 60x$		
$+5$	$-10x^2 - 10x + 60$		
	$-10x^2 - 10x + 60$		
	0		

$$x^4 + x^3 - 9x^2 - 3x + 18 = \underbrace{\frac{1}{10}(x + 1)}_{q_1(x)} \underbrace{(10x^3 - 70x + 60)}_{g_1(x)} + \underbrace{(-2x^2 - 2x - 12)}_{r_2(x)}$$

so that $q_1(x) = \frac{1}{10}(x + 1)$ and $r_2(x) = -2x^2 - 2x - 12$. Similarly,

$$10x^3 - 70x + 60 = \underbrace{(-5x + 5)}_{q_2(x)} \underbrace{(-2x^2 - 2x + 12)}_{r_2(x)} + 0$$

so that $q_2(x) = -5x + 5$ and $r_3(x) = 0$ and the process finishes in 3 steps. The last non-zero remainder is $r_2(x) = -2x^2 - 2x + 12$ whose degree is 2 (and $r_3(x) \equiv 0$). Since the GCD of two polynomials must be a monic polynomial, we divide the remainder by -2 to obtain

$$\text{GCD}(x^5 + 6x^2 - 49x + 42, x^4 + x^3 - 9x^2 - 3x + 18) = x^2 + x - 6. \qquad //$$

The above calculation can be rewritten as follows (Table 5.2):

Example 118. Find the GCD($x^4 + 3x - 2, 2x^3 + x^2 + x + 6$).

Solution. Using the Division Algorithm multiple times (as in the previous example),

$$x^4 + 3x - 2 = \left(\frac{1}{2}x - \frac{1}{4}\right)(2x^3 + x^2 + x + 6) + \left(-\frac{1}{4}x^2 + \frac{1}{4}x - \frac{1}{2}\right)$$
$$(5.3.1)$$

$$2x^3 + x^2 + x - 6 = (-8x - 12)\left(-\frac{1}{4}x^2 + \frac{1}{4}x - \frac{1}{2}\right) + 0$$

Then $r_n(x) = -\frac{1}{4}x^2 + \frac{1}{4}x - \frac{1}{2}$ and hence GCD($x^4 + 3x - 2, 2x^3 + x^2 + x + 6$) = $x^2 - x + 2$. $\qquad //$

Again the above calculation can be rewritten in table form as follows:

$\frac{1}{2}x$	$x^4+0x^3+0x^2+3x-2$	$2x^3+x^2+x+6$	$-8x$
	$x^4+\frac{1}{2}x^3+\frac{1}{2}x^2+3x$	$2x^3-2x^2+4x$	
$\frac{1}{4}$	$-\frac{1}{2}x^3-\frac{1}{2}x^2 \qquad -2$	$3x^2-3x+6$	-12
	$-\frac{1}{2}x^3-\frac{1}{4}x^2-\frac{1}{4}x-\frac{3}{2}$	$3x^2-3x+6$	
	$-\frac{1}{4}x^2+\frac{1}{4}x-\frac{1}{2}$	0	

Hence $\mathrm{GCD}(x^4+3x-2,\,2x^3+x^2+x+6) = x^2-x+2$ since $-\frac{1}{4}x^2+\frac{1}{4}x-\frac{1}{2}$ is not monic. In order to avoid arithmetic involving fractions, we can do the calculation in the following way:

x	$x^4+0x^3+0x^2+3x-2$	$2x^3+x^2+x+6$	$-2x$
	$2x^4+0x^3+0x^2+6x-4$	$2x^3-2x^2+4x$	
	$2x^4+x^3+x^2+6x$		
-1	$-x^3-x^2 \qquad -4$	$3x^2-3x+6$	-3
	$-2x^3-2x^2 \qquad -8$	$3x^2-3x+6$	
	$-2x^3-x^2-x-6$		
	x^2-x+2	0	

Exercise 72. Find the following $\mathrm{GCD}(f(x),g(x))$ if

(a) $f(x) = x^4+x^3-x^2+x-2$ and $g(x) = x^3-1$.
(b) $f(x) = x^4+x^3-x^2+x-2$, $g(x) = 3x^4-x^3-x^2-x$.

Exercise 73. Let $c \in \mathbb{R}$, $f(x) = 2x^3-4x^2+2x+(2c+4)$, and $g(x) = 3x^3-6x^2+(3c+5)$. If $\mathrm{GCD}(f(x),g(x))$ is a linear polynomial, then what is the value of c?

There was an easier way in Chap. 2 of computing the GCD of the integers a and b. That approach involved proving that there are integers x_0 and y_0 such that $\mathrm{GCD}(a,b) = x_0a + y_0b$. That result was called the Euclidean Byproduct Theorem (for integers) and is Theorem 8. Between the integers and polynomials, we now show that a similar type of result works for the GCD of two polynomials.

Theorem 23 (Euclidean Byproduct for Polynomials). *If $f(x), g(x) \in \mathbb{R}[x]$ and are not zero, then there are real polynomials $a(x)$ and $b(x)$ such that*

$$GCD(f(x), g(x)) = a(x)f(x) + b(x)g(x)$$

Example 119. Find polynomials $a(x)$ and $b(x)$ such that

$$GCD(x^4 + 3x - 2, 2x^3 + x^2 + x + 6) = a(x)\underbrace{(x^4 + 3x - 2)}_{f(x)} + b(x)\underbrace{(2x^3 + x^2 + x + 6)}_{g(x)}.$$

$$(5.3.2)$$

Solution. From Example 118, we know that $GCD(f(x), g(x)) = x^2 - x + 2$. Equation (5.3.2) is identical to

$$x^4 + 3x - 2 = \left(\frac{1}{2}x - \frac{1}{4}\right)(2x^3 + x^2 + x + 6) + \left(-\frac{1}{4}x^2 + \frac{1}{4}x - \frac{1}{2}\right).$$

Thus, we have

$$4(x^4 + 3x - 2) = (2x - 1)(2x^3 + x^2 + x + 6) + (-x^2 + x - 2) \text{ or}$$

$$x^2 - x + 2 = -4\underbrace{(x^4 + 3x - 2)}_{f(x)} + (2x - 1)\underbrace{(2x^3 + x^2 + x + 6)}_{g(x)}.$$

Hence $a(x) = -4$ and $b(x) = 2x - 1$. //

Exercise 74. Continuing Exercise 72A, find $a(x), b(x) \in \mathbb{R}[x]$, such that

$$GCD(x^4 + x^3 + 2x^2 + x + 1, x^3 - 1) = a(x)(x^4 + x^3 + 2x^2 + x + 1) + b(x)(x^3 - 1).$$

Recall that for integers a, b, and d, if $d|a$ and $d|b$ then $d|(ka + \ell b)$ for any $k, \ell \in \mathbb{Z}$. (Exercise 2, Sect. 1.1). The statement which follows (Proposition 21) is the analogy in polynomials to the statement for integers just mentioned. Its proof is left as Problem 2 of Problem Set 5.3.

Proposition 21. *If $d(x)$, $a(x)$, and $b(x)$ are non-zero polynomials with $d(x)|a(x)$ and $d(x)|b(x)$, then $d(x)|k(x)a(x) + \ell(x)b(x)$ for any polynomials $k(x)$ and $\ell(x)$.*

Example 120. Let $f(x) = ax^2 + bx + 1$ and $g(x) = bx^2 + ax + 1$. If the $GCD(f(x), g(x))$ is a linear polynomial, what is $a + b$?

Solution. The key to our method is to consider $f(x) - g(x)$. By virtue of Proposition 21 with $k(x) = 1$ and $\ell(x) = -1$, the $GCD(f, g)$ must divide $f(x) - g(x)$. But

$$f(x) - g(x) = (a - b)x^2 + (b - a)x = (a - b)x(x - 1).$$

Because we've assumed that the $GCD(f(x), g(x))$ is linear, it is either x or $x - 1$.

We now show that $GCD(f(x), g(x)) \neq x$. If x divides $f(x) = ax^2 + bx + 1$ then there is a $q(x) \in \mathbb{R}[x]$ with

$$f(x) = ax^2 + bx + 1 = xq(x).$$

But then $f(0) = 1 = 0q(0) = 0$, which is impossible.

Thus the $GCD(f(x), g(x)) = x - 1$. But, since $GCD(f(x), g(x))| f(x)$ (and $g(x)$ also) there is a polynomial $p(x)$ with

$$ax^2 + bx + 1 = (x - 1)p(x).$$

Evaluating the above when $x = 1$, we have $a + b + 1 = 0$ which gives the desired result of -1.   //

Example 121. If $f(x) = x^{50} - 2x^2 - 1$ and $g(x) = x^{48} - 3x^2 - 4$, find $GCD(f(x), g(x))$.

Solution. If $d(x)$ is a common divisor of $f(x)$ and $g(x)$, then Proposition 21 shows that $d(x)|(f(x) - x^2 g(x))$. However

$$f(x) - x^2 g(x) = (-2x^2 - 1) - (-3x^4 - 4x^2)$$
$$= 3x^4 + 2x^2 - 1 = (3x^2 - 1)(x^2 + 1).$$

Thus any common divisor, and in particular the greatest common divisor, must divide $(3x^2 - 1)(x^2 + 1)$ and so the $GCD(f(x), g(x))$ is either $\frac{1}{3}(3x^2 - 1)$ or $x^2 + 1$.

We show that the first possibility doesn't work. If $\frac{1}{3}(3x^2 - 1)| f(x)$ then $f\left(\frac{1}{\sqrt{3}}\right)$ must be zero. However,

$$f\left(\frac{1}{\sqrt{3}}\right) = \left(\frac{1}{\sqrt{3}}\right)^{50} - 2\left(\frac{1}{3}\right) - 1 = \left(\frac{1}{3}\right)^{25} - \frac{2}{3} - 1,$$

which is certainly not zero. Thus the $GCD(f(x), g(x)) = x^2 + 1$.   //

Definition. Let $f(x)$ and $g(x)$ be two polynomials over $\mathbb{R}$. The monic polynomial $c(x)$ is said the be the LEAST COMMON MULTIPLE of $f(x)$ and $g(x)$ if

(i) $f(x)|c(x)$, $g(x)|c(x)$, and
(ii) if for any $e(x) \in \mathbb{R}[x]$, whenever $f(x)|e(x)$, $g(x)|e(x)$, then $c(x)|e(x)$.

The least common multiple (LCM) for $f(x)$ and $g(x)$ will be denoted by $[f(x), g(x)] = LCM(f(x), g(x))$ and $(f(x), g(x))$ will be used for the $GCD(f(x), g(x))$.

Proposition 22. *If $f(x)$ and $g(x)$ are non-zero polynomials then*

$$f(x)g(x) = LCM(f(x), g(x))GCD(f(x), g(x)).$$

Proof. Let $d(x) = GCD(f(x), g(x))$. Then we have, since $d(x)$ must divide both $f(x)$ and $g(x)$, polynomials $k_1(x)$ and $k_2(x)$ such that

$$f(x) = d(x)k_1(x)$$

$$g(x) = d(x)k_2(x) \text{ where } GCD(k_1(x), k_2(x)) = 1$$

From the previous equations, $f(x)|d(x)k_1(x)k_2(x)$ and $g(x)|d(x)k_1(x)k_2(x)$.

Therefore $LCM(f(x), g(x)) = d(x)k_1(x)k_2(x) = \frac{f(x) \cdot g(x)}{d(x)} = \frac{f(x) \cdot g(x)}{GCD(f(x), g(x))}$
and

$$f(x) \cdot g(x) = LCM(f(x), g(x))GCD(f(x), g(x)). \qquad \blacksquare$$

The above equation can be use to find $LCM(f(x), g(x))$ if $f(x)$ and $g(x)$ are given. Namely, we find $GCD(f(x), g(x))$ first.

Example 122. Find $LCM(x^4 + 3x - 2, 2x^3 + x^2 + x + 6)$.

Solution. From Example 118, we know $GCD(x^4 + 3x - 2, 2x^3 + x^2 + x + 6) = x^2 - x + 2$. Hence

$$LCM(x^4 + 3x - 2, 2x^3 + x^2 + x + 6) = \frac{(x^4 + 3x - 2)(2x^3 + x^2 + x + 6)}{x^2 - x + 2}$$

$$= (x^4 + 3x - 2)(2x + 3). \qquad //$$

Exercise 75. Find $LCM(f(x), g(x))$ if $f(x) = 4x^4 - 4x^3 + 5x^2 - 4x + 1$ and $g(x) = 8x^3 - 6x^2 + 5x - 2$ using Proposition 22.

Problem Set 5.3:

1. Find $GCD(f(x), g(x))$ and $LCM(f(x), g(x))$ if

   (a) $f(x) = x^3 + 4x^2 - 8x + 24$ and $g(x) = x^5 - x^4 + 8x^2 - 8x$.
   (b) $f(x) = x^4 + x^3 - 9x^2 - 3x + 18$ and $g(x) = x^5 + 6x^2 - 49x + 42$.

2. Prove Proposition 21.

3. Let $f(x) = x^2 + ax + bc$ and $g(x) = x^2 + bx + ac$, and let $GCD(f(x), g(x)) = x - d$. Find $LCM(f(x), g(x))$.

4. Let $f(x) = x^3 - 2x^2 + ax - 3$, $g(x) = x^2 - bx + 2$, where $a, b, c \in \mathbb{N}$, and let $GCD(f(x), g(x)) = x - c$. Find $LCM(f(x), g(x))$.

5. Let $f(x) = x^2 + ax + b$ and $g(x) = x^2 + bx + a$ where $a, b \in \mathbb{R}$. If the $GCD(f(x), g(x))$ is linear, find $LCM(f(x), g(x))$.

Chapter 6
Matrices and Systems of Linear Equations

Introduction This chapter discusses the solutions of a system of linear equations by viewing the solutions as a set and interpreting that set geometrically. Matrices and operations on matrices are briefly reviewed and their effect on the related solution sets is also described geometrically.

6.1 Matrix Operations

Overview The key conceptual point and reason that matrices are so important is the relationship between matrices and the solution set of a system of linear equations which, will be discussed in the next section. Because matrices and their operations are introduced in school mathematics (for example, in Algebra 2 courses in high schools in the United States), we will not provide much detail in this section.

We will pay attention mostly to the cases of 2 or 3 dimensional Euclidean space (also called 2-space or 3-space) or equivalently, 2 or 3 equations in 2 or 3 unknowns because the basic ideas of the subject are in 2 and 3 space. Everything extends easily to higher dimensions but that generality is one taken in a college level course "linear algebra". Restricting to 2 or 3 dimensions saves some formalism and much notation which allows the underlying ideas of the subject to be clearer.

Definition. An $(m \times n)$ MATRIX is a collection of real numbers arranged in m rows and n columns. The SIZE of such a matrix is $m \times n$. The $m \times n$ ZERO MATRIX is one in which all entries are zero. We write $\mathbf{0}$ for the zero matrix of any size.

We will write a capital letter, say A, for the matrix. The ith row and jth column entry of A is denoted by a_{ij}.

© Springer International Publishing Switzerland 2015
R.S. Millman et al., *Problems and Proofs in Numbers and Algebra*,
DOI 10.1007/978-3-319-14427-6_6

Definition. Two matrices are EQUAL if they have the same size and their corresponding entries are the same. In symbols, two $m \times n$ matrices A and B are equal if

$$a_{ij} = b_{ij} \text{ for all } 1 \leq i \leq m \text{ and } 1 \leq j \leq n.$$

A matrix, A, can be MULTIPLIED BY A REAL NUMBER λ by multiplying each of its entries by λ. More formally we have the following definition of scalar multiplication.

Definition. If $A = [a_{ij}]$ is an $m \times n$ matrix and λ is a real number, then the SCALAR MULTIPLICATION λA is the $m \times n$ matrix with entries defined by the (i, j) entry of λA is λa_{ij}, or

$$(\lambda A)_{ij} = \lambda a_{ij} \text{ for all } 1 \leq i \leq m \text{ and } 1 \leq j \leq n.$$

The resulting matrix, λA, is the matrix whose (i, j) component is $\lambda a_{i,j}$. In other words, the components of λA are exactly the components of A multiplied by λ.

Definition. In symbols, if A and B are two $m \times n$ matrices then their SUM $C = A + B$ is the matrix with entries defined by

$$c_{ij} = a_{ij} + b_{ij} \text{ for all } 1 \leq i \leq m \text{ and } 1 \leq j \leq n.$$

If A and B are matrices of the same size then SUBTRACTION of B from A is defined to be $A + (-1)B = A - B$.

The matrix addition and scalar multiplication can be viewed as component wise operations. This reduces the matrix operations to performing addition or multiplication of real numbers.

Exercise 76. For any matrix A, prove that the zero matrix $\mathbf{0}$ is the additive identity and $-A = (-1)A$ is the additive inverse of A. More precisely, for any matrix A, prove that:

1. $A + \mathbf{0} = A$
2. $A + (-A) = \mathbf{0}$.

This exercise can be done easily component-wise.

Definition. If A is $n \times n$ then its DIAGONAL elements are a_{ii} where $i = 1, \ldots, n$. A matrix A is called a DIAGONAL MATRIX if all non-diagonal entries are zero.

One very special and useful diagonal matrix is the identity, I.

Definition. The $n \times n$ matrix I in which each element on the diagonal is 1 and all others are zero is called the MULTIPLICATIVE IDENTITY or, more simply, the IDENTITY. Note that I can be written component-wise

$$I_{ij} = \begin{cases} 1 & \text{if } i = j \\ 0 & \text{if } i \neq j \end{cases}$$

where $1 \leq i \leq n$ and $1 \leq j \leq n$.

To define matrix multiplication, it is convenient to first introduce a symbol to show the sum of a certain finite set of numbers $\{x_k\}$.

Definition. If x_k is a real number for each integer k between ℓ and n inclusive, then the symbol $\sum\limits_{\ell}^{n} x_k$ is the SUM of $\{x_k\}$ for k between ℓ and n inclusive, or

$$\sum_{\ell}^{n} x_k = x_\ell + x_{\ell+1} + \cdots + x_{n-1} + x_n.$$

The left hand side of the previous equation is called "summation notation".

Exercise 77. If $x_k = k$ show that

$$\sum_{k=1}^{n} x_k = \frac{n(n+1)}{2}$$

by using Gauss's insight about arithmetic progressions.

Exercise 78. Looking at a high school algebra text, use summation notation, and arithmetic and geometric progressions to express the following sums.

A. The sum of an arithmetic progression whose first term is a and whose common difference is d.
B. The sum of a geometric progression whose first term is a and whose common ratio is r. Assume that $r \neq 1$.

We now turn to the definition of matrix multiplication.

Definition. If A and B are matrices in which the number of columns of A is the same as the number of rows of B, then the PRODUCT of A and B, AB, is the matrix $C = (c_{ij})$ where

$$c_{ij} = \sum_{k=1}^{p} a_{ik} b_{kj}.$$

Note that if A is $m \times p$ and B is $p \times n$ then $C = AB$ is $m \times n$. Furthermore, these formulae in the 2×2 case are

$$\begin{bmatrix} a_{11} & a_{12} \\ a_{21} & a_{22} \end{bmatrix} \begin{bmatrix} b_{11} & b_{12} \\ b_{21} & b_{22} \end{bmatrix} = \begin{bmatrix} a_{11}b_{11} + a_{12}b_{21} & a_{11}b_{12} + a_{12}b_{22} \\ a_{21}b_{11} + a_{22}b_{21} & a_{21}b_{12} + a_{22}b_{22} \end{bmatrix} \tag{6.1.1}$$

which is the usual result.

It is important to note that unlike matrix addition or scalar multiplication of real numbers, matrix multiplication can not be viewed as a component wise operation and it is not necessarily commutative (if $n > 1$); that is, it may be that $AB \neq BA$ for certain matrices A and B.

Exercise 79.

A. If A is an $n \times n$ matrix, prove that $IA = A$ and $AI = A$.
B. Show that matrix multiplication is not commutative. More precisely, find 2×2 matrices, A and B such that $AB \neq BA$.

The inverse of a matrix is very helpful in applications to numerical analysis, number theory, geometry, and has many uses outside of mathematics. It can be thought of as similar to taking the inverse of a real number. Just as nearly all real numbers a have an inverse (in fact, for all $a \neq 0$), so do inverses exist for most but not all matrices.

Definition. The $n \times n$ matrix A has a MULTIPLICATIVE INVERSE, written as A^{-1}, if there exists a matrix A^{-1} such that

$$AA^{-1} = A^{-1}A = I.$$

If A has a multiplicative inverse, A is called INVERTIBLE.

Exercise 80. If $A = \begin{bmatrix} 6 & 2 \\ 10 & 4 \end{bmatrix}$ show that

$$A^{-1} = \begin{bmatrix} 1 & -\frac{1}{2} \\ -\frac{5}{2} & \frac{3}{2} \end{bmatrix}.$$

We will now associate to each matrix A a number, the determinant of A (written $\det A$) which will give us a precise criterion for a matrix to have an inverse and also give a formula (Theorem 24) for the inverse of A. Knowing which matrices have inverses and how to compute the inverse is a subject of great importance in higher mathematics and its applications.

Definition. If A is the 2×2 matrix

$$A = \begin{bmatrix} a & b \\ c & d \end{bmatrix}$$

then the DETERMINANT of A is the number

$$\det A = \begin{vmatrix} a & b \\ c & d \end{vmatrix} = ad - bc.$$

If A is the 3×3 matrix

$$A = \begin{bmatrix} a_{11} & a_{12} & a_{13} \\ a_{21} & a_{22} & a_{23} \\ a_{31} & a_{32} & a_{33} \end{bmatrix}$$

then

$$\det A = \begin{vmatrix} a_{11} & a_{12} & a_{13} \\ a_{21} & a_{22} & a_{23} \\ a_{31} & a_{32} & a_{33} \end{vmatrix} = a_{11}a_{22}a_{33} + a_{12}a_{23}a_{31} + a_{13}a_{21}a_{32}$$

$$- (a_{11}a_{23}a_{32} + a_{12}a_{21}a_{33} + a_{13}a_{22}a_{31}). \qquad (6.1.2)$$

The comments of this paragraph sketch the general definition of the determinant and can be skipped on the first reading of the text. The above definition is valid for $n \times n$ matrices when $n = 2$ or 3 only. A careful look at the 3×3 case hints at what the definition of $\det A$ should be if $n > 3$. Note the first step in the computation shows that $\det A$ is obtained by going down the first column and multiplying the (i, j) coefficient by the determinant of a 2×2 matrix which is obtained by crossing out the i-th row and the j-th column. The $\det A$ is obtained by computing the alternating sum of those numbers, i.e.

$$\det A = a_{11} \det \begin{bmatrix} a_{22} & a_{23} \\ a_{32} & a_{33} \end{bmatrix} - a_{12} \det \begin{bmatrix} a_{21} & a_{23} \\ a_{31} & a_{33} \end{bmatrix} + a_{13} \det \begin{bmatrix} a_{21} & a_{22} \\ a_{31} & a_{32} \end{bmatrix}$$

$$= (a_{11}a_{22}a_{33} + a_{12}a_{23}a_{31} + a_{13}a_{21}a_{32}) - (a_{11}a_{23}a_{32} + a_{12}a_{21}a_{33} + a_{13}a_{22}a_{31})$$

The last two sentences show how one would generalize the definition of determinants to the 4×4 case and beyond. Since we only need the 2×2 and 3×3 cases in this text, we will not go beyond $n = 3$. Furthermore, since this book is not a text in linear or matrix algebra, we will only use the determinant to compute inverses in Theorem 24 and state Cramer's Rule in Corollary 9 and Sect. 6.4. You have already seen Cramer's Rule in high school. Here we are making careful definitions showing that Cramer's Rule actually works for $n = 2$ and $n = 3$.

Remark. There are similar expression for a 3×3 determinant by expanding by any row or column.

Exercise 81. Compute the determinant of the following matrices.

1. $\begin{bmatrix} 3 & -2 \\ 1 & 4 \end{bmatrix}$

2. $\begin{bmatrix} 1 & -2 & 4 \\ 1 & 2 & -2 \\ 0 & -1 & 3 \end{bmatrix}$

3. $\begin{bmatrix} 1 & 2 & 3 \\ 4 & 5 & 6 \\ 7 & 8 & 9 \end{bmatrix}$

The next proposition is especially useful when trying to calculate the determinant of a matrix. The proof in the case of $n = 2$ is left as exercise. The 3×3 case can be done the same way (which is sometimes called "by brute force") although in advanced courses easier ways are used. We will not ask for a proof when $n = 3$ or higher.

Proposition 23. *If A and B are both $n \times n$ then $\det(AB) = \det A \det B$.*

Exercise 82. Prove Proposition 23 for $n = 2$.

Definition. If A is an $m \times n$ matrix then A^{T}, called the TRANSPOSE of A is the $n \times m$ matrix whose (i, j) entry is the (j, i) entry of A.

As an example, note that if

$$A = \begin{bmatrix} 1 & 2 & 3 \\ -3 & 4 & 5 \end{bmatrix}$$

then

$$A^{\mathrm{T}} = \begin{bmatrix} 1 & -3 \\ 2 & 4 \\ 3 & 5 \end{bmatrix}.$$

If A is square, then A^{T} merely reflects the matrix across its diagonal, or, said another way, A^{T} = "flipping A across its diagonal". The next proposition shows that the transpose of a product is the product of the transpose in the opposite order. (Remember, order matters for matrix multiplication!)

Proposition 24. *If A_1 and A_2 are both square matrices of size $n \times n$ when $n = 2$ or 3, then*

$$(A_1 A_2)^{\mathrm{T}} = A_2{}^{\mathrm{T}} A_1{}^{\mathrm{T}}.$$

Proof. In the 2×2 case, we let

$$A_1 = \begin{bmatrix} a_1 & b_1 \\ c_1 & d_1 \end{bmatrix}$$

$$A_2 = \begin{bmatrix} a_2 & b_2 \\ c_2 & d_2 \end{bmatrix}$$

then

$$A_1{}^{\mathrm{T}} = \begin{bmatrix} a_1 & c_1 \\ b_1 & d_1 \end{bmatrix} \text{ and } A_2{}^{\mathrm{T}} = \begin{bmatrix} a_2 & c_2 \\ b_2 & d_2 \end{bmatrix}.$$

So being careful of the order:

$$A_2{}^{\mathrm{T}} A_1{}^{\mathrm{T}} = \begin{bmatrix} a_2 a_1 + c_2 b_1 & a_2 c_1 + c_2 d_1 \\ b_2 a_1 + d_2 b_1 & b_2 c_1 + d_1 d_2 \end{bmatrix}$$

and

$$A_1 A_2 = \begin{bmatrix} a_1 a_2 + b_1 c_2 & a_1 b_2 + b_1 d_2 \\ c_1 a_2 + d_1 c_2 & b_2 c_1 + d_1 d_2 \end{bmatrix}.$$

Thus $(A_1 A_2)^{\mathrm{T}} = A_2{}^{\mathrm{T}} A_1$.

The proof of the case when $n = 3$ follows along the same lines (by brute force), but needs summation notation and matrix entries written with their usual double subscripts. ∎

In order to write out a formula for the inverse of a 3×3 matrix (when one exists) we first need the definitions of cofactors and the minors of a matrix.

Definition. Let (i, j) be a fixed pair of integers and write

$$A = \begin{bmatrix} a_{11} & a_{12} & a_{13} \\ a_{21} & a_{22} & a_{23} \\ a_{31} & a_{32} & a_{33} \end{bmatrix}.$$

Then the (i, j) minor of A, denoted M_{ij} is the determinant of the matrix formed by crossing out the ith row and j^{th} column of A. The COFACTOR c_{ij} of A is given by $c_{ij} = (-1)^{i+j} M_{ij}$.

Example 123. Let

$$A = \begin{bmatrix} 2 & 3 & 1 \\ 0 & 4 & 2 \\ 1 & 2 & -1 \end{bmatrix}.$$

Find the cofactors, c_{11}, c_{12}, c_{23}.

Solution. To find the minor M_{11}, we delete the first row and first column of A and evaluate the determinant of the resulting matrix:

$$A = \begin{bmatrix} 2 & 3 & 1 \\ 0 & 4 & 2 \\ 1 & 2 & -1 \end{bmatrix}, M_{11} = \begin{vmatrix} 4 & 2 \\ 2 & -1 \end{vmatrix} = -4 - 4 = -8$$

Then the cofactor $c_{11} = (-1)^{1+1} M_{11} = -8$. Similarly, we have:

$$c_{12} = (-1)^{1+2} M_{12} = -\begin{vmatrix} 0 & 2 \\ 1 & -1 \end{vmatrix} = 2,$$

$$c_{23} = (-1)^{2+3} M_{23} = -\begin{vmatrix} 2 & 3 \\ 1 & 2 \end{vmatrix} = -1.$$

//

Exercise 83. Find all cofactors of the matrix A from Example 123. (There are a total of 9, can you see why?)

We now have the background to write out the inverse of a matrix.

Theorem 24. *If A is a 2×2 or 3×3 matrix then*

(i) *A is invertible if and only if $\det A \neq 0$.*
(ii) *(2×2 case) If $A = \begin{bmatrix} a_1 & b_1 \\ a_2 & b_2 \end{bmatrix}$ and $\det A \neq 0$ then*

$$A^{-1} = \frac{1}{\det A} \begin{bmatrix} b_2 & -b_1 \\ -a_2 & a_1 \end{bmatrix}. \tag{6.1.3}$$

(iii) *(3×3 cases) If $\det A \neq 0$ and c_{ij} is the (i, j) cofactor of A, then*

$$A^{-1} = \frac{1}{\det A} \begin{bmatrix} c_{11} & c_{12} & c_{13} \\ c_{21} & c_{22} & c_{23} \\ c_{31} & c_{32} & c_{33} \end{bmatrix}^{T}. \tag{6.1.4}$$

Proof. We first note that if A has an inverse, say $A^{-1} = X$, then $AX = I$. Thus, from Proposition 23

$$(\det A)(\det X) = \det I = 1$$

and so $\det A \neq 0$. To finish Part (i), we must now show, in the case in which $n = 2$ or 3, that $\det A \neq 0$ implies that A is invertible.

In fact, to show that A is invertible we can just write down a formula for A^{-1}. Equations (6.1.3) and (6.1.4) give exactly such formulas so we need only show that the right hand side of those equations when multiplied by A give I. We will now do the case when $n = 2$.

$$A\frac{1}{\det A}\begin{bmatrix} b_2 & -b_1 \\ -a_2 & a_1 \end{bmatrix} = \frac{1}{\det A}\begin{bmatrix} a_1 & b_1 \\ a_2 & b_2 \end{bmatrix}\begin{bmatrix} b_2 & -b_1 \\ -a_2 & a_1 \end{bmatrix}$$

$$= \frac{1}{\det A}\begin{bmatrix} a_1b_2 - b_1a_2 & 0 \\ 0 & a_1b_2 - b_1a_2 \end{bmatrix} = I$$

The same computation in the opposite order gives $A^{-1}A = I$ also, which completes the case when $n = 2$.

The case in which $n = 3$ gets bogged down in notation and we will omit it as it is just a straightforward computation (by brute force). ∎

Example 124. Find A^{-1} by using Theorem 24 where

$$A = \begin{bmatrix} 5 & -4 & 1 \\ -4 & 6 & 2 \\ 1 & -4 & 3 \end{bmatrix}.$$

Solution. Here we will compute the nine elements of the cofactor matrices of A:

$$c_{11} = (-1)^{1+1}\det\begin{bmatrix} 6 & 2 \\ -4 & 3 \end{bmatrix} = 26,$$

$$c_{12} = (-1)^{1+2}\det\begin{bmatrix} -4 & 2 \\ 1 & 3 \end{bmatrix} = 14,$$

$$c_{13} = (-1)^{1+3}\det\begin{bmatrix} -4 & 6 \\ 1 & -4 \end{bmatrix} = 10,$$

$$c_{21} = (-1)^{2+1}\det\begin{bmatrix} -4 & 1 \\ -4 & 3 \end{bmatrix} = 8,$$

$$c_{22} = (-1)^{2+2}\det\begin{bmatrix} 5 & 1 \\ 1 & 3 \end{bmatrix} = 14,$$

$$c_{23} = (-1)^{2+3} \det \begin{bmatrix} 5 & -4 \\ 1 & -4 \end{bmatrix} = 16,$$

$$c_{31} = (-1)^{3+1} \det \begin{bmatrix} -4 & 1 \\ 6 & 2 \end{bmatrix} = -14,$$

$$c_{32} = (-1)^{3+2} \det \begin{bmatrix} 5 & 1 \\ -4 & 2 \end{bmatrix} = -14,$$

$$c_{33} = (-1)^{3+3} \det \begin{bmatrix} 5 & -4 \\ -4 & 6 \end{bmatrix} = 14.$$

$$\det A = a_{11}c_{11} + a_{12}c_{12} + a_{13}c_{13} = 5 \cdot 26 - 4 \cdot 14 + 1 \cdot 10 = 84.$$

Hence, according to Theorem 24,

$$A^{-1} = \frac{1}{84} \begin{bmatrix} 26 & 14 & 10 \\ 8 & 14 & 10 \\ -14 & -14 & 14 \end{bmatrix}^{\mathrm{T}} = \frac{1}{84} \begin{bmatrix} 26 & 8 & -14 \\ 14 & 14 & -14 \\ 10 & 16 & 14 \end{bmatrix}.$$

//

Problem Set 6.1:

1. Find the sum or product where possible.

   (a) $\begin{bmatrix} 3 & 1 \\ 4 & 2 \end{bmatrix} \begin{bmatrix} 2 & 2 \\ -2 & 1 \\ -1 & 2 \end{bmatrix}$

   (b) $\begin{bmatrix} -3 & 4 \end{bmatrix} \begin{bmatrix} 7 & 7 & 7 \\ 3 & 2 & 1 \\ 5 & 6 & 11 \end{bmatrix}$

   (c) $2 \begin{bmatrix} 3 & 1 & 4 \end{bmatrix} + \begin{bmatrix} 1 & 2 & -1 \end{bmatrix} \begin{bmatrix} 1 & 0 & 2 \\ 0 & 0 & 3 \\ -1 & 0 & 4 \end{bmatrix}$

   (d) $\begin{bmatrix} 3 & 1 \\ -1 & 2 \end{bmatrix}^2 - 5 \begin{bmatrix} 3 & 1 \\ -1 & 2 \end{bmatrix} + 6 \begin{bmatrix} 1 & 0 \\ 0 & 1 \end{bmatrix}$

2. If $A = \begin{bmatrix} 4 & 2 & 1 \\ 6 & 1 & -5 \end{bmatrix}$, find A^{T} and A^{-1} if they exist.

3. Prove each of the following statements either in the $n = 2$ case or $n = 3$ depending on your instructor's wish. All matrices are square of size $n \times n$.

   (a) $\lambda(B + C) = \lambda B + \lambda C, \lambda \in \mathbb{R}$
   (b) $(\lambda B)C = \lambda(BC), \lambda \in \mathbb{R}$
   (c) $A(B + C) = AB + AC$

4. Give an example to show, in the $n = 2$ case, that for $n \times n$ matrices A, B, C, and D it is possible that

   (i) $AB \neq BA$
   (ii) $CD = 0$ but $C \neq 0$ and $D \neq 0$. (Of course, this situation never occurs for real numbers.)

5. Find the inverse of each of the following matrices or explain why there is no inverse

   (a) $A = \begin{bmatrix} 6 & 2 \\ 1 & 4 \end{bmatrix}$

   (b) $B = \begin{bmatrix} -2 & 1 \\ 2 & 1 \end{bmatrix}$

   (c) $C = \begin{bmatrix} 6 & 3 & 1 \\ 4 & 2 & 7 \\ 10 & 5 & 8 \end{bmatrix}$

   (d) $D = \begin{bmatrix} 1 & 0 & 1 \\ 1 & 1 & 0 \\ 0 & 1 & 1 \end{bmatrix}$

   (e) $E = \begin{bmatrix} 2 & 3 & 0 \\ 7 & 8 & 9 \\ -4 & -6 & 1 \end{bmatrix}$

6. By first computing the minors of the matrix A of Exercise 83, find A^{-1}.

7. (a) If A and B are both not invertible square matrices, show that $A + B$ is not invertible or give a counterexample to this statement in the case of $n = 2$ or 3.
   (b) If A and B are both invertible square matrices, show that $A + B$ is invertible or give a counterexample to this statement in the case of $n = 2$ or 3.
   (c) If C is an invertible matrix and $\lambda \in \mathbb{R}$ is not zero, show that λC is invertible or give a counterexample to this statement in the case of $n = 2$ or 3.

8. Let A and B be $n \times n$ matrices.

   (a) Prove or disprove $\det(A + B) = \det A + \det B$ in the case when (i) $n = 2$ (ii) $n = 3$
   (b) Prove or disprove that for $\lambda \in \mathbb{R}$,

   $$\det(\lambda A) = \lambda^n \det A$$

   in the case where (i) $n = 2$ (ii) $n = 3$

9. Let $A = \begin{bmatrix} -1 & 0 & 8 \\ 2 & 2 & 0 \\ -1 & 0 & -4 \end{bmatrix}$, $B = \begin{bmatrix} -1 & 4 & 8 \\ 1 & -4 & -6 \\ 0 & 3 & 1 \end{bmatrix}$.

   (a) Determine the determinants of the matrices A and B.

   (b) Determine the determinants of the matrices AB and BA.

   (c) Prove or Disprove: In the case of all 3×3 matrices, $\det(AB) = \det(A)\det(B)$.

10. It is known that $\begin{bmatrix} a & 4 \\ -1 & 2 \end{bmatrix}^{-1}$ does not exist. What is the value of a?

6.2   Systems of Linear Equations in the Plane

Overview In this section, we will deal first with two linear equations in two unknowns describing the various solutions that are possible. Of course, the word linear means "line" and that gives us a geometric picture of the algebra that we will be doing. Throughout this section, we will interpret our results in terms of the geometry of straight lines in the plane.

Let's think about the possibilities for two lines in the plane. Geometrically, there are only three choices for the relative position of two lines in the plane. They are either identical, intersect at exactly one point, or don't intersect at all. We can also see these results from the viewpoint of algebra. If we have two linear equations in two unknowns, there are also only three choices: the equations have identical solutions (the lines they define are actually the same line), have a unique point as a simultaneous solution (the lines have exactly one common point), or have no solution at all (the lines do not intersect). (See Table 6.1, page 164, for a summary of these interpretations). First, some definitions are needed.

Definition. A STRAIGHT LINE in the plane is an equation of the form $ax + by = c$ where a and b are not both zero. Alternatively, a straight line defined by the real numbers a, b, and c is given as a set

$$L = \{(x, y) \mid ax + by = c\}$$

where a and b are not both zero.

From your school mathematics, it is common to think of a straight line as an equation of the form $y = mx + k$ where m is the slope and k is constant (the usual letter b is being reserved for another constant). Geometrically, because the line goes through $(0, k)$, k is called the y-INTERCEPT, or more precisely, the y-coordinate of the y-intercept. Since we will also need to talk about vertical lines, we will define a straight line as above.

Looking at the definition of a straight line above, the case in which $b = 0$ but $a \neq 0$ is that of a vertical line. If $a = 0$ but $b \neq 0$ then the line is horizontal. More generally, if $b \neq 0$ then the line L is not vertical and

$$y = mx + k \text{ where } m = -\frac{a}{b} \text{ is the slope and } k = \frac{c}{a}.$$

Instead of looking at the situation geometrically, we now view it algebraically as a set of equations. It is important to note that it is not the equations themselves which matter, but the set of points that satisfy them.

Definition. The SOLUTION SET for the n equations in 2 variables $f_1(x, y), \ldots,$ $f_n(x, y)$ is

$$\{(x, y) \in \mathbb{R}^2 \mid f_1(x, y) = f_2(x, y) = \cdots = f_n(x, y) = 0\}.$$

The system is LINEAR if each of the functions $f_1(x, y), f_2(x, y), \ldots, f_n(x.y)$ is a straight line.

Many people make the common error of confusing an equation with its solution set. For example, $x + y = 0$ and $2x + 2y = 0$ are not the same equation but they have the same solution set $\{(x, y) \mid y = -x\}$, the line through the origin with slope -1. We can double the first equation to get the second which changes the first but does not change its solution set. The idea of changing equations without changing the solution set is very important and we'll see it algebraically, geometrically, and through matrix operations

The condition that "not both a and b are zero" is said more simply in mathematical notation as $a^2 + b^2 \neq 0$. This will come in handy when we are involved in more than one equation.

We will now work with two linear equations in two unknowns, show the geometry of the situation and how trying to find intersection points (a solution set) leads naturally to the algebraic concept of the determinant.

Definition. We will consider the case of the two lines

$$L_1 : a_1 x + b_1 y = c_1 \qquad\qquad a_1^2 + b_1^2 \neq 0$$
$$L_2 : a_2 x + b_2 y = c_2 \qquad\qquad a_2^2 + b_2^2 \neq 0 \qquad\qquad (6.2.1)$$

The system (6.2.1) which is determined by L_1 and L_2 is called HOMOGENEOUS if $c_1 = c_2 = 0$ and it is NON-HOMOGENEOUS otherwise. The matrix of coefficients of the system (6.2.1) is $A = \begin{bmatrix} a_1 & b_1 \\ a_2 & b_2 \end{bmatrix}$.

Since a solution to system (6.2.1) means that a point $P = (x, y)$ is on both lines L_1 and L_2, P is an intersection point of the lines L_1 and L_2. Looking at things geometrically, there are three cases, which contain the definition of a system of equations as consistent, inconsistent, dependent and independent.

Possibility 1. The lines L_1 and L_2 intersect. The only alternatives are that L_1 and L_2 may intersect in exactly one point or their solution set is the same straight line so that there are infinitely many points of intersection; that is, $L_1 = L_2$. In this case, the system (6.2.1) is called CONSISTENT.

Possibility 2. The lines L_1 and L_2 do not intersect. The system of L_1 and L_2 (6.2.1) will then have no solution and the system is called INCONSISTENT.

Consider the possibility where the lines L_1 and L_2 intersect. In this case, the system (6.2.1) is said to be INDEPENDENT if it has exactly one solution or no solutions. If the system is not independent, it is called DEPENDENT. We will see that if (6.2.1) is dependent then the first equation becomes a consequence of the second equation (or vice versa). Here is a summary of what we will prove in this section.

Theorem 25. *Consider the system of two equations in two unknowns given by Eq. (6.2.1). The only possibilities for solutions sets are given by Table 6.1.*

Proof. For convenience, we first let A to be the 2×2 matrix of coefficients of (6.2.1) with d its determinant:

$$A = \begin{bmatrix} a_1 & b_1 \\ a_2 & b_2 \end{bmatrix}, \qquad \det A = a_1 b_2 - b_1 a_2 = d.$$

Going back to System (6.2.1), we multiply the first equation of (6.2.1) by b_2 and the second by b_1 giving

$$a_1 b_2 x + b_1 b_2 y = c_1 b_2$$

and

$$a_2 b_1 x + b_1 b_2 y = c_2 b_1.$$

Table 6.1 Character of possible solutions to a linear system of two equations in two unknowns

Case	# of solutions	Solution set	System dependent/ independent	System consistent/ inconsitent
1	None	Empty	Independent	Inconsistent
2	1	Point	Independent	Consistent
3	Infinitely many	Line	Dependent	Consistent

Subtracting these two equations and recognizing that

$$a_1b_2 - b_1a_2 = \det A = d$$

gives,

$$dx = c_1b_2 - b_1c_2. \qquad (6.2.2)$$

Similarly, multiplying the first equation by a_2 and the second by a_1 and subtracting gives

$$dy = a_1c_2 - a_2c_1 \qquad (6.2.3)$$

Thus (x, y) is in the solution set of Eq. (6.2.1) exactly when (6.2.2) and (6.2.3) are satisfied. We now have three cases.

Case 1. $(d \neq 0)$ By simply dividing (6.2.2) and (6.2.3) by d we obtain

$$x = \frac{\det\begin{bmatrix} c_1 & b_1 \\ c_2 & b_2 \end{bmatrix}}{d}, \quad y = \frac{\det\begin{bmatrix} a_1 & c_1 \\ a_2 & c_2 \end{bmatrix}}{d} \text{ if } d = \det A = \det\begin{bmatrix} a_1 & b_1 \\ a_2 & b_2 \end{bmatrix} \neq 0. \qquad (6.2.4)$$

Case 2. If $d = 0$ and the numerators $\det\begin{bmatrix} c_1 & b_1 \\ c_2 & b_2 \end{bmatrix}$ and $\det\begin{bmatrix} a_1 & c_1 \\ a_2 & c_2 \end{bmatrix}$, both equal 0 and then the system (6.2.1) has infinitely many solutions.

Case 3. If $d = 0$, $\det\begin{bmatrix} c_1 & b_1 \\ c_2 & b_2 \end{bmatrix} \neq 0$, and $\det\begin{bmatrix} a_1 & c_1 \\ a_2 & c_2 \end{bmatrix} \neq 0$, then the system (6.2.1) has no solutions.

Together, these three cases give the Table 6.1.   ∎

Corollary 9 (Cramer's Rule). *Consider the system of the two equations in two unknowns (6.2.1) whose matrix of coefficients is $A = \begin{bmatrix} a_1 & b_1 \\ a_2 & b_2 \end{bmatrix}$ and whose constants c_1, c_2 are on the righthand side of the equations. If $d = \det A \neq 0$ then there is a unique solution to the system and the values for (x, y) are*

$$x = \frac{\det\begin{bmatrix} c_1 & b_1 \\ c_2 & b_2 \end{bmatrix}}{d} \text{ and } y = \frac{\det\begin{bmatrix} a_1 & c_1 \\ a_2 & c_2 \end{bmatrix}}{d}. \qquad (6.2.5)$$

Thus, the condition that $\det A \neq 0$ means that the solution set of the system (6.2.1) consists of a single point whose coordinates are given by (6.2.5).

Note that if the homogeneous system

$$a_1 x + b_1 y = 0$$
$$a_2 x + b_2 y = 0$$

has non-trivial solutions (that is, solutions other than $x = y = 0$) then

$$d = \det \begin{bmatrix} a_1 & b_1 \\ a_2 & b_2 \end{bmatrix} = 0.$$

Note also that the Case 1, Case 2 and Case 3 can be simplified as follows. Let

$$a_1 x + b_1 y = c_1,$$
$$a_2 x + b_2 y = c_2.$$

Then

Case 1. If $\frac{a_1}{a_2} \neq \frac{b_1}{b_2}$, then it has one solution.

Case 2. If $\frac{a_1}{a_2} = \frac{b_1}{b_2} = \frac{c_1}{c_2}$, then it has infinitely many solutions.

Case 3. If $\frac{a_1}{a_2} = \frac{b_1}{b_2} \neq \frac{c_1}{c_2}$, then it has no solution.

We will use Cramer's Rule for $n \times n$ matrices in Sect. 6.4 when $n/ \geq 3$.

Example 125. For each of the nonhomogeneous systems (a), (b), (c), what are the solutions?

(a) System (a) is

$$2x + 5y = 2$$
$$x + y = 3$$

Since $\frac{2}{1} \neq \frac{5}{1}$, it has one solution.

(b) System (b) is

$$2x + 4y = 7$$
$$x + 2y = 6$$

Since $\frac{2}{1} = \frac{4}{2} \neq \frac{7}{6}$, it has no solution.

(c) System (c) is

$$x - 3y = 5$$
$$3x - 4y = 15$$

Since $\frac{1}{3} = \frac{-3}{-9} = \frac{5}{15}$, it has infinitely many solutions.

Example 126. Find the solution set for

$$3x - 8y = 5$$
$$4x + 9y = 11.$$

Solution. The solution set uses Cramer's Rule. Since $\det \begin{bmatrix} 3 & -8 \\ 4 & 9 \end{bmatrix} = 59$, $\det \begin{bmatrix} 5 & -8 \\ 11 & 9 \end{bmatrix} = 133$, and $\det \begin{bmatrix} 3 & 5 \\ 4 & 11 \end{bmatrix} = 13$, Eq. (6.2.5) shows that $x = \frac{133}{59}$ and $y = \frac{13}{59}$. The solution set is the one point $\left(\frac{133}{59}, \frac{13}{59} \right)$.   //

We now give some examples of systems of equations which do not look like they are problems about linear systems. It is a transformation which allows us to solve the problems using linear systems.

Example 127. Find all solutions of the system of equations

$$3x + y = 6xy \text{ and } 2y - x = 9xy \qquad (6.2.6)$$

Solution. We are being asked to find the solution set of (6.2.6). Although it doesn't look like a linear system, we can use the principles already established to solve it. Firstly, it is obvious that $(0, 0)$ is a solution. Furthermore, if either $x = 0$ or $y = 0$, then so must be the other one. Thus we are looking for solutions in which neither x nor y is zero. We will use a transformation whose variables are now X and Y with $x \neq 0$ and $y \neq 0$.

$$X = \tfrac{1}{x}, \quad Y = \tfrac{1}{y} \qquad (6.2.7)$$

and Eq. (6.2.6) becomes

$$X + 3Y = 6$$
$$2X - Y = 9. \qquad (6.2.8)$$

Equation (6.2.4) gives $X = \frac{33}{7}$ and $Y = \frac{3}{7}$ as a solution to (6.2.8). Since $X = \frac{1}{x}$ and $Y = \frac{1}{y}$, the solutions set for (6.2.6) is

$$\left\{ (0, 0), \left(\tfrac{7}{33}, \tfrac{7}{3} \right) \right\}.$$

//

Note that examining the case where x or y is zero allows us to make the transformation (or change of variables) given by Eq. (6.2.7).

Example 128. Solve the system

$$\frac{3}{x-3} + \frac{4}{2y-1} = 7$$

and   (6.2.9)

$$\frac{4}{x-3} - \frac{3}{2y-1} = 1.$$

Solution. Since $x \neq 3$ and $y \neq \frac{1}{2}$ (because Eq. (6.2.9) will make no sense), we may make the transformations $X = \frac{1}{x-3}$ and $Y = \frac{1}{2y-1}$. Then we have a system of equations in our new variables X and Y which is

$$3X + 4Y = 7$$
$$4X - 3Y = 1.$$ (6.2.10)

Now, d, which is the determinant of the coefficient matrix is given by

$$\det \begin{bmatrix} 3 & 4 \\ 4 & -3 \end{bmatrix} = -25.$$

So Corollary 9 shows that

$$X = \frac{\det \begin{bmatrix} 7 & 4 \\ 1 & -3 \end{bmatrix}}{-25} = \frac{-21 - 4}{-25} = 1$$

and

$$Y = \frac{\det \begin{bmatrix} 3 & 7 \\ 4 & 1 \end{bmatrix}}{-25} = \frac{-25}{-25} = 1.$$

We have now solved the system (6.2.10) and found X and Y. However we were asked to solve system (6.2.9) for x and y. Since $1 = X = \frac{1}{x-3}$, we see $x = 4$. Similarly, $1 = Y = \frac{1}{2y-1}$ so that $y = 1$ and $(4,1)$ is the unique solution to Example 128.   //

Example 129. Let $a \in \mathbb{R}$. Discuss the characteristics of the solution set in terms of a of the system

$$ax - 6y = 5a - 3$$
$$2x + (a - 7)y = 29 - 7a.$$

Solution. Here, we are asked how the solution set varies as a does. To use Theorem 24 (and Corollary 9), we need to compute three determinants. The coefficient matrix of the system is

$$\begin{bmatrix} a & -6 \\ 2 & a-7 \end{bmatrix}.$$

Thus

$$d = \det \begin{bmatrix} a & -6 \\ 2 & a-7 \end{bmatrix} = a^2 - 7a + 12$$

$$\det \begin{bmatrix} 5a-3 & -6 \\ 29-7a & a-7 \end{bmatrix} = 5a^2 - 80a + 195$$

and

$$\det \begin{bmatrix} a & 5a-3 \\ 2 & 29-7a \end{bmatrix} = -7a^2 + 19a + 6.$$

Putting these last three equations into (6.2.2) and (6.2.3) shows that, for a solution (x, y) to exist,

$$(a^2 - 7a + 12)x = 5a^2 - 80a + 195$$

and

$$(a^2 - 7a + 12)y = -7a^2 + 19a + 6.$$

Factoring each side of the last two equations shows that x must satisfy

$$(a-3)(a-4)x = 5(a-3)(a-13) \tag{6.2.11}$$

and y must satisfy

$$(a-3)(a-4)y = -(7a+2)(a-3). \tag{6.2.12}$$

If $a = 3$ then (6.2.11) and (6.2.12) both reduce to $0 = 0$ and there is an infinite solution set (given by either $3x - 6y = 12$ or, what is the same $2x - 4y = 8$).
If $a \neq 3$ then (6.2.11) and (6.2.12) become

$$(a-4)x = 5(a-13) \text{ and } (a-4)y = -(7a+2). \tag{6.2.13}$$

Table 6.2 Solutions to
Example 129

Value of a	Solution set
$a = 3$	$\{(x, y) \mid 2x - 4y = 8\}$
$a = 4$	Empty
$a \neq 3, a \neq 4$	$\left\{\left(\frac{5(a-13)}{a-4}, -\frac{7a+2}{a-4}\right)\right\}$

If $a = 4$ then (6.2.13) shows that there is no solution. If $a \neq 4$, then we can solve
(6.2.13) directly and come up with

$$x = \frac{5(a - 13)}{a - 4} \text{ and } y = -\frac{7a + 2}{a - 4}$$

as the unique solution. We collect these solutions in the table above (Table 6.2).

//

Problem Set 6.2:

For each of the Problems 1 through 4, give all solutions or state that there are
no solutions. State whether the system is dependant, independent, consistent or
inconsistent. Justify your answers.

1.

$$6x - 4y = 4$$
$$-3x + 2y = \pi$$

2.

$$10x - 6y = 8$$
$$-35x - 15y = 20$$

3.

$$4y + z = 6$$
$$-y + 2z = 9$$

4.

$$x + 3z = 8$$
$$4x - z = 9$$

5. Let x and z be real numbers. Find all solutions to the equations

$$4x - z = 2xz$$

and

$$2z - 2x = 7xz.$$

6. Are there real numbers w and t with

$$3t + 7w = 2wt \text{ and } 2t + 4w = 7wt?$$

7. Solve if possible for w and t.

$$(2(w - 1) + t = 2t(w - 1)$$

and

$$3w + 3t = 4t(w - 1).$$

8. Solve for x and y, if possible.

$$\frac{2}{x + 5} - \frac{3}{y - 7} = 4$$

and

$$\frac{3}{x + 5} - \frac{2}{y - 7} = 6$$

9. Solve for x and y if possible.

$$2^x = 5 \cdot 2^y + 7$$
$$3 \cdot 2^x + 4 \cdot 2^y = 2$$

10. Solve for x and y, if possible.

$$e^x + 2e^y = 4$$
$$-3e^x + 4e^y = 5$$

11. Solve for x and y, if possible.

$$e^{x+2} - e^y = 3$$
$$3e^x + 4e^y = 4$$

12. Solve for x and y if possible.

$$e^{2x} + 2^{y+1} = 7$$
$$e^{2x} - 2^y = 5$$

13. Let $a \in R$. Discuss the characteristic of the solution set of the system.

$$(a - 3)x - 2y = 2a$$
$$3x + (2a + 1)y = -a - 2$$

14. Solve for x and y, given real numbers a and b.

$$x + ay + a^2 = 0$$
$$x + by + b^2 = 0$$

6.3   Systems of Linear Equations in Euclidean n-Space

Overview The many similar computations and row operations in the first two sections of Chap. 6 lead us to consider if there is a natural connection between the material. In this section, we show how operations on matrices can be interpreted in terms of solutions sets to a system of linear equations. Cramer's Rule (Sect. 6.4) follows easily from the use of matrices. We will go on to use the ideas of Sect. 6.5 to simplify computations such as finding the GCD of two integers.

We will continue to use the notation of Eq. (6.2.1) by starting with two lines L_1 and L_2, which are each given by two equations

$$L_1: a_1 x + b_1 y = c_1 \qquad a_1^2 + b_1^2 \neq 0$$
$$L_2: a_2 x + b_2 y = c_2 \qquad a_2^2 + b_2^2 \neq 0 \qquad (6.3.1)$$

and ask for the solutions set to these equations but this time matrices are used. Once again, we write

$$A = \begin{bmatrix} a_1 \ b_1 \\ a_2 \ b_2 \end{bmatrix} \text{ and } d = \det A$$

for the matrix of coefficients of (6.3.1). Let $C = \begin{bmatrix} c_1 \\ c_2 \end{bmatrix}$. Then the system (6.3.1) becomes

$$A \begin{bmatrix} x \\ y \end{bmatrix} = C. \tag{6.3.2}$$

What is the inverse of A? In order for that to happen $d = \det A \neq 0$. In that case, Eq. (6.1.4) shows that

$$A^{-1} = \frac{1}{d} \begin{bmatrix} b_2 & -b_1 \\ -a_2 & a_1 \end{bmatrix}. \tag{6.3.3}$$

Proposition 25. *If the matrix of coefficients of the system (6.3.1) has a non-zero determinant then there is a unique solution to (6.3.1) $\begin{bmatrix} x \\ y \end{bmatrix}$ such that*

$$\begin{bmatrix} x \\ y \end{bmatrix} = A^{-1}C.$$

and

$$x = \frac{\det \begin{bmatrix} c_1 & b_1 \\ c_2 & b_2 \end{bmatrix}}{\det \begin{bmatrix} a_1 & b_1 \\ a_2 & b_2 \end{bmatrix}} \tag{6.3.4}$$

$$y = \frac{\det \begin{bmatrix} a_1 & c_1 \\ a_2 & c_2 \end{bmatrix}}{\det \begin{bmatrix} a_1 & b_1 \\ a_2 & b_2 \end{bmatrix}}. \tag{6.3.5}$$

Proof. Let $X = \begin{bmatrix} x \\ y \end{bmatrix}$ be the 2×1 matrix of unknowns. Since $AX = C$ and A has an inverse, $A^{-1}AX = A^{-1}C$ so that $\begin{bmatrix} x \\ y \end{bmatrix} = A^{-1}C$. Now we will actually compute x and y using (6.3.3).

$$\begin{bmatrix} x \\ y \end{bmatrix} = A^{-1}C = \frac{1}{d} \begin{bmatrix} b_2 & -b_1 \\ -a_2 & a_1 \end{bmatrix} \begin{bmatrix} c_1 \\ c_2 \end{bmatrix}$$

$$= \frac{1}{d} \begin{bmatrix} c_1 b_2 - b_1 c_2 \\ a_1 c_2 - a_2 c_1 \end{bmatrix}.$$

The entry $c_1 b_2 - b_1 c_2$ is

$$\det \begin{bmatrix} c_1 & b_1 \\ c_2 & b_2 \end{bmatrix}$$

whereas the entry $c_1b_2 - b_1c_2$ is

$$\det\begin{bmatrix} a_1 & b_1 \\ a_2 & b_2 \end{bmatrix},$$

which results in (6.3.4) and (6.3.5).   ∎

Notice once again, when the appropriate machinery is derived, how easy some results appear. The solution (when the matrix of coefficients does not have zero determinant) to the system (6.2.1) is given by Eqs. (6.3.4) and (6.3.5).

Is there an easy way to remember the solution of the linear system $AX = C$ if $d = det(A) \neq 0$? Yes, there is! Examining the conclusion of Proposition 25 shows that

$$x = \frac{b_2c_1 - b_1c_2}{d}$$

is the same as the $det\begin{bmatrix} c_1 & b_1 \\ c_2 & b_2 \end{bmatrix}$. The matrix is obtained by taking out the first column of A (the "x column" if you like) and replacing it with the coefficient column $C = \begin{bmatrix} c_1 \\ c_2 \end{bmatrix}$. The solution for y is the $det\begin{bmatrix} a_1 & c_1 \\ a_2 & c_2 \end{bmatrix} / d$ which is the determinant of the matrix found by replacing the "y column" by the coefficient matrix C. Thus, the way to remember Eq. (6.3.4) is to replace the x-row in A with the constants of the original equation, compute the determinant and divide by $d = det(A)$. A similar statement holds for Eq. (6.3.5).

The use of matrices also allows us to handle systems of m equations in n unknowns. This is really the subject of a semester course in linear algebra and will not be treated in full generality here.

However, we will now look at the situation in which the number of equations is the same as the number of variables, ultimately specializing to the case of $n = 3$ dimensions (or a system of 3 linear equations in 3 variables).

Definition. A LINEAR EQUATION in n variables $x_1, \ldots, x_n$ is an equation of the form

$$a_1x_1 + a_2x_2 + \cdots + a_nx_n = c$$

where $a_1, \ldots, a_n$ are real numbers, not all zero and $c \in \mathbb{R}$.

A LINEAR SYSTEM of n equations in n unknowns is a collection of n linear equations in n variables.

There are three points to note. The first is that if a linear system of n equations in n unknowns is given then there are n^2 real numbers a_{ij} (called the COEFFICIENTS OF THE SYSTEM) and n real numbers $c_1, \ldots, c_n$ such that

$$a_{11}x_1 + a_{12}x_2 + \cdots + a_{1n}x_n = c_1$$

$$a_{21}x_1 + a_{22}x_2 + \cdots + a_{2n}x_n = c_2$$

$$\vdots \qquad \qquad \vdots \qquad\qquad\qquad (6.3.6)$$

$$a_{n1}x_1 + a_{n2}x_2 + \cdots + a_{nn}x_n = c_n$$

and for each i, $\sum_{j=0}^{n} a_{ij}^2 \neq 0$ (this is the statement of the technical point that, for a fixed i, not all a_{ij} are zero).

The second is the 3×3 case. When there are 3 equations in 3 unknowns, (6.3.6) is written (usually with $x_1 = x$, $x_2 = y$, and $x_3 = z$) as

$$a_{11}x + a_{12}y + a_{13}z = c_1$$

$$a_{21}x + a_{22}y + a_{23}z = c_2 \qquad\qquad (6.3.7)$$

$$a_{31}x + a_{32}y + a_{33}z = c_3.$$

Equation (6.3.7) should be familiar to you from Algebra II when you were solving 3 linear equations in 3 unknowns.

The third point is that the number of equations and the number of unknowns is the same, n. Solving for 2 equations in 3 unknowns (or 3 equations is 2 unknowns) gives quite different alternatives than 3 equations in 3 unknowns.

Proposition 26. *The system of linear equations (6.3.6) has a unique solution if the determinant of the matrix of coefficients of the system is not zero.*

Proof. Let A be the matrix of coefficients of the system, so that

$$A = \begin{bmatrix} a_{11} & a_{12} & \cdots & a_{1n} \\ a_{21} & a_{22} & \cdots & a_{2n} \\ \vdots & \vdots & \ddots & \vdots \\ a_{n1} & a_{n2} & \cdots & a_{nn} \end{bmatrix}$$

and let C be the $n \times 1$ matrix given by the right hand side of the system (6.3.6)

$$C = \begin{bmatrix} c_1 \\ \vdots \\ c_n \end{bmatrix}.$$

If we write the unknowns as an $n \times 1$ matrix, X, where

$$X = \begin{bmatrix} x_1 \\ \vdots \\ x_n \end{bmatrix}$$

then (6.3.6) becomes

$$AX = C. \tag{6.3.8}$$

Since $\det A \neq 0$, there is an inverse to A and multiplying the last equation by A^{-1} yields $A^{-1}AX = A^{-1}C$ or

$$X = A^{-1}C.$$

Of course, we could substitute $X = A^{-1}C$ into Eq. (6.3.8) and see that X satisfies that equation.

We have, from Eq. (6.1.4) a precise form for the inverse of a matrix, so that, if $c_{i,j}$ is the (i, j) cofactor of A, then

$$X = \frac{1}{\det A} \begin{bmatrix} c_{11} & c_{12} & \cdots & c_{13} \\ c_{21} & c_{22} & \cdots & c_{23} \\ \vdots & \vdots & \ddots & \vdots \\ c_{31} & c_{32} & \cdots & c_{33} \end{bmatrix}^{\mathrm{T}} C.$$

In the 3×3 case

$$X = \begin{bmatrix} x \\ y \\ z \end{bmatrix} = \frac{1}{\det A} \begin{bmatrix} c_{11} & c_{12} & c_{13} \\ c_{21} & c_{22} & c_{23} \\ c_{31} & c_{32} & c_{33} \end{bmatrix}^{\mathrm{T}} \begin{bmatrix} c_1 \\ c_2 \\ c_3 \end{bmatrix}. \tag{6.3.9}$$

∎

We now consider ELEMENTARY ROW OPERATIONS on matrices which will be very useful in the next section. The key point is that performing an elementary row operation on a matrix does not change the solution set to the corresponding linear system (Theorem 26).

Definition. Let B be an $m \times n$ matrix. An ELEMENTARY ROW OPERATION on B is one of

(i) Two rows of B are interchanged
(ii) Multiply the ith row by a real number $c \neq 0$.
(iii) To change row i, row number j is multiplied by a real number, λ, then added to row i and placed in row i. This means that $R_i + \lambda R_j \rightarrow R_i$. The other rows of the original matrix B stay the same.

Example 130.

(a) If $B = \begin{bmatrix} 2 & 3 & 1 \\ 4 & 6 & 3 \end{bmatrix}$, then what is the result of multiplying the first row of B by (-2) and adding it to row 2?

(b) If the elementary row operation, interchange the second and third rows of

$$D = \begin{bmatrix} 2\,3 & 5\,4 \\ 3\,1 & 7\,2 \\ -3\,3 & -4\,7 \end{bmatrix},$$

what is the resulting matrix?

Solution. The elementary row operation adds $(-4\ -6\ -2)$ to the second row but leaves the first row unchanged. Thus

$$B \longrightarrow \begin{bmatrix} 2\,3\,1 \\ 0\,0\,1 \end{bmatrix}. \tag{6.3.10}$$

For the second part, interchanging the second and third rows gives

$$D \longrightarrow \begin{bmatrix} 2\,3 & 5\,4 \\ -3\,3 & -4\,7 \\ 3\,1 & 7\,2 \end{bmatrix}. \tag{6.3.11}$$

$/\!/$

For motivation as to why we use elementary row operations, let's look at the linear systems that B or D represent and see what happens to the solution set under an elementary row operation.

Consider the linear system

$$2x + 3y = 1$$
$$4x + 6y = 3 \tag{6.3.12}$$

The coefficient matrix of the system is $\begin{bmatrix} 2\,3 \\ 4\,6 \end{bmatrix}$ and the two constants $\begin{bmatrix} 1 \\ 3 \end{bmatrix}$ give the matrix B of Example 130(a):

$$B = \begin{bmatrix} 2\,3 & 1 \\ 4\,6 & 3 \end{bmatrix}.$$

Multiplying the first equation in the system (6.3.12) by (-2) and adding it to the second equation of (6.3.12) gives (repeating the first equation)

$$2x + 3y = 1$$
$$0x + 0y = 1. \tag{6.3.13}$$

This last system is represented by

$$\begin{bmatrix} 2 & 3 & 1 \\ 0 & 0 & 1 \end{bmatrix}$$

(6.3.14)

but we've not changed the solution of (6.3.12) because we have performed simple algebra (multiplying one equation by (-2) and adding it to the other).

Exercise 84. What system does the matrix D of Example 130 represent? What system does the matrix D after the elementary row operations described in the example represent? Do the two systems have the same solutions set? What does the second row mean in Eq. (6.3.14)?

Remark. We have just observed that elementary row operations allow us to manipulate a system of represents them without changing the solution set of the original equation.

Thus we have the following theorem.

Theorem 26. *If S is a linear system represented by the matrix B and $e(B)$ is the matrix which comes from applying an elementary row operation to B, then B and $e(B)$ are matrices whose systems have the same solution set.*

Example 131. Find the solutions of the system of Eq. (6.3.12) [see Example 130(a)].

Solution. Equation (6.3.12) is represented by $B = \begin{bmatrix} 2 & 3 & 1 \\ 4 & 6 & 3 \end{bmatrix}$ and it has the same solution set as in (6.3.14)

$$\begin{bmatrix} 2 & 3 & 1 \\ 0 & 0 & 1 \end{bmatrix}.$$

But $e(B)$ represents the solutions to the equation

$$2x + 3y = 1$$
$$0x + 0y = 1 \text{ or } 0 = 1.$$

Thus, there are no solutions the system of (6.3.12).   //

Example 132. Solve the linear system

$$y - 3z = 4$$
$$2x + 3y - z = 6$$
$$4x + 5y + 3z = 8$$

by using elementary row operations.

Solution.

$$\begin{bmatrix} 0 & 1 & -3 & 4 \\ 2 & 3 & -1 & 6 \\ 4 & 5 & 3 & 8 \end{bmatrix} \rightarrow \begin{bmatrix} 2 & 3 & -1 & 6 \\ 0 & 1 & -3 & 4 \\ 4 & 5 & 3 & 8 \end{bmatrix} \quad \text{interchange row 1 and row 2}$$

$$\rightarrow \begin{bmatrix} 2 & 3 & -1 & 6 \\ 0 & 1 & -3 & 4 \\ 0 & -1 & 5 & -4 \end{bmatrix} \quad \text{Add } -2 \text{ times row 1 to row 3}$$

$$\rightarrow \begin{bmatrix} 2 & 3 & -1 & 6 \\ 0 & 1 & -3 & 4 \\ 0 & 0 & 2 & 0 \end{bmatrix} \quad \text{Add row 2 to row 3}$$

$$\rightarrow \begin{bmatrix} 2 & 3 & -1 & 6 \\ 0 & 1 & -3 & 4 \\ 0 & 0 & 1 & 0 \end{bmatrix} \quad \text{Multiply row 3 by } \frac{1}{2}. \text{ This shows } z = 0$$

$$\rightarrow \begin{bmatrix} 2 & 3 & 0 & 6 \\ 0 & 1 & -3 & 4 \\ 0 & 0 & 1 & 0 \end{bmatrix} \quad \text{Add row 3 to row 1}$$

$$\rightarrow \begin{bmatrix} 2 & 3 & 0 & 6 \\ 0 & 1 & 0 & 4 \\ 0 & 0 & 1 & 0 \end{bmatrix} \quad \text{Add 3 times row 3 to row 2, this shows } y = 4$$

$$\rightarrow \begin{bmatrix} 2 & 0 & 0 & -6 \\ 0 & 1 & 0 & 4 \\ 0 & 0 & 1 & 0 \end{bmatrix} \quad \text{Add } -3 \text{ times row 2 to row 1}$$

$$\rightarrow \begin{bmatrix} 1 & 0 & 0 & -3 \\ 0 & 1 & 0 & 4 \\ 0 & 0 & 1 & 0 \end{bmatrix} \quad \text{Multiply row 1 by } \frac{1}{2}. \text{ This shows } x = -3$$

Hence we have found the solutions $x = -3, y = 4, z = 0$. Note the above process is called Gaussian elimination, after the German mathematician C. F. Gauss (1777–1855).   //

Problem Set 6.3:

1. Let $A = \begin{bmatrix} 1 & 4 & 3 \\ 2 & -3 & 5 \\ -2 & 3 & 1 \end{bmatrix}$.

   (a) What is the resulting matrix if you multiply the first row of A by 3 and then add it to row 2?

(b) What is the resulting matrix if you interchange the second and third rows of A?

(c) What is the resulting matrix if you perform both operations from Parts (a) and (b)? Does the final matrix depend on which order you perform the operations?

2. Find solutions to the system of equations below using determinants:

$$2x + 4y = 12$$
$$-x + 5y = 9$$
(6.3.15)

3. Find solutions to the system of equations below using determinants:

$$x + y + z = 3$$
$$2x + 3y - z = -4$$
$$-4x - 2y + 3z = 6$$
(6.3.16)

4. Can you perform elementary row operations on the matrices below and end up with the matrix I. If so, list the necessary row operations and show each resulting matrix after performing the operation. If it is impossible, explain why.

(a) $\begin{bmatrix} 1 & 0 & 0 \\ 4 & 2 & 0 \\ 0 & -3 & 1 \end{bmatrix}$.

(b) $\begin{bmatrix} 1 & -2 & 0 \\ 0 & -2 & 0 \\ 0 & -2 & 1 \end{bmatrix}$.

(c) $\begin{bmatrix} 1 & 1 & 0 \\ 0 & 0 & 3 \\ 0 & 0 & 1 \end{bmatrix}$.

(d) $\begin{bmatrix} 2 & 4 & 3 \\ 0 & -2 & 4 \\ 1 & -1 & 4 \end{bmatrix}$. $\left(\text{Answer:} \begin{bmatrix} -2/7 & -5/14 & 11/7 \\ 2/7 & 5/14 & -4/7 \\ -1/7 & 3/7 & -2/7 \end{bmatrix} \right)$.

5. Determine if the below augmented matrices represent systems with one, infinitely many, or no solutions first by solving for the determinant and then again using Gaussian elimination but do not actually solve the system. In each case, determine which method was simpler and why.

i. $\begin{bmatrix} 3 & 3 & | & 2 \\ 4 & 1 & | & 3 \end{bmatrix}$

ii. $\begin{bmatrix} 2 & -1 & | & 5 \\ -4 & 2 & | & 0 \end{bmatrix}$

iii. $\begin{bmatrix} -5 & 3 & | & 1 \\ 10 & -6 & | & -2 \end{bmatrix}$

iv. $\begin{bmatrix} 2\,4 & 12 & | & 18 \\ 0\,6 & -9 & | & 18 \\ 0\,0 & 17 & | & 204 \end{bmatrix}$

6. Solve the following systems of equations by Gaussian elimination.

   i.

$$x + 3y + z = 4$$
$$2x + 8y - 2z = 2$$
$$3x + 9y - 9z = 16$$

   ii.

$$2x + 4y - 7z = 6$$
$$x - 3y - 2z = 2$$
$$6x + 12y - 21z = 24$$

   iii.

$$x + 3z = 6$$
$$6y - 2z = 2$$
$$x - 12y + 7z = 16$$

6.4 System of Linear Equations: Cramer's Rule

Overview The Cramer's Rule in Corollary 9 can be generalized easily to a system of n linear equations in n variables. In Sect. 6.4, we will use $n = 3$ throughout and follow the use of matrices.

Theorem 27. *Consider system of 3 linear equations in 3 unknowns*

$$a_{11}x + a_{12}y + a_{13}z = b_1, \tag{6.4.1}$$

$$a_{21}x + a_{22}y + a_{23}z = b_2, \tag{6.4.2}$$

$$a_{31}x + a_{32}y + a_{33}z = b_3. \tag{6.4.3}$$

We write

$$A = \begin{bmatrix} a_{11} & a_{12} & a_{13} \\ a_{21} & a_{22} & a_{23} \\ a_{31} & a_{32} & a_{33} \end{bmatrix} \quad \text{and} \quad \Delta = \det A = |A|$$

$$\Delta_x = \begin{vmatrix} b_1 & a_{12} & a_{13} \\ b_2 & a_{22} & a_{23} \\ b_3 & a_{32} & a_{33} \end{vmatrix}, \quad \Delta_y = \begin{vmatrix} a_{11} & b_1 & a_{13} \\ a_{21} & b_2 & a_{23} \\ a_{31} & b_3 & a_{33} \end{vmatrix}, \quad \Delta_z = \begin{vmatrix} a_{11} & a_{12} & b_1 \\ a_{21} & a_{22} & b_2 \\ a_{31} & a_{32} & b_3 \end{vmatrix}.$$

If $\Delta \neq 0$, then

$$x = \frac{\Delta_x}{\Delta}, y = \frac{\Delta_y}{\Delta}, z = \frac{\Delta_z}{\Delta}.$$

Proof. We know that Eqs. (6.4.1), (6.4.2), and (6.4.3) can be written as

$$\begin{bmatrix} a_{11} & a_{12} & a_{13} \\ a_{21} & a_{22} & a_{23} \\ a_{31} & a_{32} & a_{33} \end{bmatrix} = \begin{bmatrix} x \\ y \\ z \end{bmatrix} = \begin{bmatrix} b_1 \\ b_2 \\ b_3 \end{bmatrix}$$

Theorem 22(iii) on page 159 uses the (i, j) cofactors of A whose notation is c_{ij} for $1 \leq i \leq 3$ and $1 \leq j \leq 3$. Thus we have

$$AX = \begin{bmatrix} b_1 \\ b_2 \\ b_3 \end{bmatrix} \quad \text{where } X = \begin{bmatrix} x \\ y \\ z \end{bmatrix}.$$

The Eq. (6.1.4) shows that

$$X = \begin{bmatrix} x \\ y \\ z \end{bmatrix} = \frac{1}{|A|} \begin{bmatrix} c_{11} & c_{12} & c_{13} \\ c_{21} & c_{22} & c_{23} \\ c_{31} & c_{32} & c_{33} \end{bmatrix}^{\text{T}} \begin{bmatrix} b_1 \\ b_2 \\ b_3 \end{bmatrix} = \frac{1}{|A|} \begin{bmatrix} c_{11} & c_{21} & c_{31} \\ c_{12} & c_{22} & c_{32} \\ c_{13} & c_{23} & c_{33} \end{bmatrix} \begin{bmatrix} b_1 \\ b_2 \\ b_3 \end{bmatrix}$$

Therefore, with Δ and $|A|$ being the same notation for $\det A$,

$$x = \frac{1}{|A|}(b_1 c_{11} + b_2 c_{21} + b_3 c_{31}) = \frac{1}{|A|} \Delta_x,$$

$$y = \frac{1}{|A|}(b_1 c_{12} + b_2 c_{22} + b_3 c_{32}) = \frac{1}{|A|} \Delta_y,$$

$$z = \frac{1}{|A|}(b_1 c_{13} + b_2 c_{23} + b_3 c_{33}) = \frac{1}{|A|} \Delta_z.$$

■

Example 133. Use Cramer's Rule to solve

$$7x - y - z = 0,$$
$$10x - 2y + z = 8,$$
$$6x + 3y - 2z = 7.$$

Solution. The determinants are

$$\Delta = \begin{vmatrix} 7 & -1 & -1 \\ 10 & -2 & 1 \\ 6 & 3 & -2 \end{vmatrix} = -61, \quad \Delta_x = \begin{vmatrix} 0 & -1 & -1 \\ 8 & -2 & 1 \\ 7 & 3 & -2 \end{vmatrix} = -61,$$

$$\Delta_y = \begin{vmatrix} 7 & 0 & -1 \\ 10 & 8 & 1 \\ 6 & 7 & -2 \end{vmatrix} = -183, \quad \Delta_z = \begin{vmatrix} 7 & -1 & 0 \\ 10 & -2 & 8 \\ 6 & 3 & 7 \end{vmatrix} = -244.$$

Thus $x = \frac{\Delta_x}{\Delta} = \frac{-61}{-61} = 1$, $y = \frac{\Delta_y}{\Delta} = \frac{-183}{-61} = 3$, $z = \frac{\Delta_z}{\Delta} = \frac{-244}{-61} = 4.$ //

Exercise 85. Use Cramer's Rule to solve

$$4x + 3y + 3z = 3,$$
$$2x - 4y + 4z = 7,$$
$$3x + 2y + 6z = 3.$$

(Answer: $x = \frac{31}{25}$, $y = -\frac{4}{5}$, $z = \frac{11}{75}$.)

Example 134. Use Cramer's Rule to solve the system of equations in x, y and z.

$$\frac{3}{y+z} + \frac{6}{z+x} + \frac{1}{x+y} = 1,$$

$$\frac{6}{y+z} + \frac{4}{z+x} + \frac{1}{x+y} = 3,$$

$$\frac{15}{y+z} + \frac{2}{z+x} + \frac{3}{x+y} = 6.$$

Solution. Let $X = \frac{1}{y+z}$, $Y = \frac{1}{z+x}$, $Z = \frac{1}{x+y}$. We now have three linear equations in three unknowns X, Y and Z.

$$3X + 6Y + Z = 1,$$
$$6X + 4Y - Z = 3,$$
$$15X + 2Y - 3Z = 6.$$

By using Cramer's Rule, we obtain

$$X = \frac{1}{6}, \quad Y = \frac{1}{4}, \quad Z = -1.$$

Thus $\frac{1}{y+z} = \frac{1}{6}, \frac{1}{z+x} = \frac{1}{4}, \frac{1}{x+y} = -1$. Therefore, we have three linear equations in x, y and z.

$$y + z = 6,$$
$$z + x = 4,$$
$$x + y = -1.$$

Solving this system, we have $x = -\frac{3}{2}, y = \frac{1}{2}, z = \frac{11}{2}$.   //

Exercise 86. By using Cramer's Rule, solve

$$\frac{1}{x} + \frac{1}{y} + \frac{1}{z} = 0,$$

$$\frac{4}{x} + \frac{3}{y} + \frac{2}{z} = 5,$$

$$\frac{3}{x} + \frac{2}{y} + \frac{4}{z} = -4.$$

(Answer: $x = 1/2, y = 1, z = -1/3$.)

Example 135. It is known that the following system has a unique solution, find the value of a.

$$ax + y + z = 1,$$
$$x + ay + z = 1,$$
$$x + y + az = 1.$$

Solution. Since

$$\Delta = \begin{vmatrix} a & 1 & 1 \\ 1 & a & 1 \\ 1 & 1 & a \end{vmatrix} = a^3 + 1 + 1 - a - a - a = a^3 - 3a + 2 = (a-1)^2(a+2),$$

we must have $a \neq 1$, $a \neq -2$. Thus, all values of a except $a = 1$ and $a = 2$ have a unique solution.   //

Exercise 87. It is known that the following system has solution $x = 0$, $y = 0$, $z = 0$ for any value of $a \in R$.

$$-x + y + z = ax,$$
$$x - y + z = ay,$$
$$x + y - z = az.$$

Find the values of a so that the system has no other solution. (Answer: $a \neq 1$ or $a \neq 2$.)

Example 136. Let 3 positive integers be given whose sum is 51. If the first positive integer is divided by the second one, then the quotient is 2 and the remainder is 5. If the second positive integer is divided by the third one, then the quotient is 3 and the remainder is 2. Find the value of these three integers.

Solution. Let x, y, z be the given positive integers. Then

$$x + y + z = 51,$$
$$x = 2y + 5,$$
$$y = 3z + 2.$$

This system is equivalent to

$$x + y + z = 51,$$
$$x - 2y = 5,$$
$$y - 3z = 2.$$

Using the same notation as Theorem 27

$$\Delta = \begin{vmatrix} 1 & 1 & 1 \\ 1 & -2 & 0 \\ 0 & 1 & -3 \end{vmatrix} = 6 + 1 + 3 = 10,$$

$$\Delta_x = \begin{vmatrix} 51 & 1 & 1 \\ 5 & -2 & 0 \\ 2 & 1 & -3 \end{vmatrix} = 51(6) + 5 + 4 + 15 = 330,$$

$$\Delta_y = \begin{vmatrix} 1 & 51 & 1 \\ 1 & 5 & 0 \\ 0 & 2 & -3 \end{vmatrix} = -15 + 2 + 153 = 140,$$

$$\Delta_z = \begin{vmatrix} 1 & 1 & 51 \\ 1 & -2 & 5 \\ 0 & 1 & 2 \end{vmatrix} = 40,$$

We have $x = \frac{\Delta_x}{\Delta} = 33$, $y = \frac{\Delta_y}{\Delta} = 14$, $z = \frac{\Delta_z}{\Delta} = 4$.   //

Exercise 88. If person A, B, C work together for one job, they needs 10 days to finish it. If B and C work together, then they can finish it in 15 days. If A works for 15 days, then C needs 30 days to finish it (B is not working). How many days will be needed if A, B, C works alone? (Answer: $x = 30$, $y = 20$, $z = 60$.)

Example 137. Let a, b, c be the lengths of a triangle, and let

$$x + y + z = 0,$$
$$ax + by + cz = 0,$$
$$a^2x + b^2y + c^2z = 0$$

have only one solution $(0,0,0)$. Then what kind of equations can this triangle be described by (if at all?)

Solution. Since

$$\Delta = \begin{vmatrix} 1 & 1 & 1 \\ a & b & c \\ a^2 & b^2 & c^2 \end{vmatrix} = (a-b)(b-c)(c-a) \neq 0,$$

we have $a \neq b \neq c$. Hence the triangle is not equilateral triangle nor isosceles triangle. Note: the matrix

$$\begin{bmatrix} 1 & 1 & 1 \\ a & b & c \\ a^2 & b^2 & c^2 \end{bmatrix}$$

is called a Vandermonde matrix, named after A. T. Vandermonde (1735–1796).   //

Exercise 89. Compute the determinants of these Vandermonde matrices:

$$\begin{vmatrix} 1 & 1 & 1 \\ 10 & 11 & 12 \\ 100 & 121 & 144 \end{vmatrix} \quad \text{and} \quad \begin{vmatrix} 1 & 5 & 25 \\ 1 & 9 & 81 \\ 1 & 12 & 144 \end{vmatrix}.$$

Example 138 (Fitting for Quadratic Polynomial Curves). Let $A = (-1, 2)$, $B = (1, 5)$, and $C = (2, 3)$. Show that $\{A, B, C\}$ forms a quadratic polynomial curve.

Solution. Let $p(x) = a + bx + cx^2$ and hence,

$$2 = p(-1) = a - b + c \quad \text{for } A$$
$$5 = p(1) = a + b + c \quad \text{for } B$$
$$3 = p(2) = a + 2b + 4c \quad \text{for } C.$$

The solution of this system of linear equations in the variables a, b, and c is

$$a = \frac{14}{3}, \quad b = \frac{3}{2}, \quad \text{and } c = -\frac{7}{6}.$$

Thus, the quadratic polynomial is $p(x) = \frac{14}{3} + \frac{3}{2}x - \frac{7}{6}x^2$.   //

Exercise 90. Find a quadratic polynomial curve that fits the points $D = (-2, 4)$, $E = (1, 5)$, and $F = (3, 7)$.

Problem Set 6.4:

For Problems 1 through 6, solve the given system of linear equations by Cramer's rule.

1.

$$x + 2y + z = 6$$
$$2x - 4y + z = 7$$
$$3x + 3y + 5z = 10$$

2.

$$3x + 4y + 3z = 0$$
$$x + 2y - z = -1$$
$$x + y + 2z = 1$$

3.

$$x + y - z = 0$$
$$x - 2y + z = -3$$
$$4x + y - 2z = -3$$

4.

$$\frac{1}{x} + \frac{1}{y} + \frac{1}{z} = 0$$

$$\frac{3}{x} - \frac{3}{y} + \frac{2}{z} = 6$$

$$\frac{4}{x} + \frac{2}{y} + \frac{3}{z} = -5$$

5.

$$\frac{3}{y+z} + \frac{6}{z+x} + \frac{1}{x+y} = 1$$

$$\frac{6}{y+z} + \frac{4}{z+x} - \frac{1}{x+y} = 3$$

$$\frac{15}{y+z} + \frac{2}{z+x} - \frac{3}{x+y} = 6$$

6.

$$x + y + z = 1$$

$$ax + by + cz = d$$

$$a^2x + b^2y + c^2z = d^2$$

7. Assume that

$$x + 3y + 2z = -3$$

$$2x + y + 3z = 1$$

$$x + ay + z = 6$$

has infinitely many solutions. Then what is the value of a.

8. Assume that

$$ax - y + 2z = 0$$

$$x + y - z = 0$$

$$3x + y = 0$$

has infinitely many solutions. Then what is the value of a.

9. Assume that the following systems of linear equations have the same solution. What is the value of a, b, c?

$$x + y + z = 6 \qquad\qquad ax - by + cz = 12$$
$$3x - y + 2z = 9 \quad \text{and} \quad ax + by + cz = 32$$
$$3x - 2y - 5z = 0 \qquad\qquad cx - by - az = 0$$

6.5   Applications of Matrix Operations to the GCD

Overview For readers who have studied matrix algebra, there is a more procedural way to find the GCD(a, b) and a particular solution of the integral linear combination. We will describe this matrix process through Example 139 on the next page which readdresses the case of Example 43 (page 54) through the use of matrices.

To accomplish the matrix method for GCDs, we first write the two integers a, b as a column $[\begin{smallmatrix} a \\ b \end{smallmatrix}]$ and then append the identity matrix. We write $N = [\begin{smallmatrix} a \\ b \end{smallmatrix}]$ and I for the identity which appears as

$$A = [\, N \, I \,],$$

a 2×3 matrix. We can now use elementary row operations on A, where A has only integers as elements. These row operations are called INTEGRAL ELEMENTARY ROW OPERATIONS. We then add or subtract an integral multiple of one row from another until the $(2, 1)$ or $(1, 1)$ entry is zero. The top row of the reduced matrix gives the results (or the bottom if the $(1, 1)$ entry is zero).

Here is a justification of this matrix approach. Note that the above elementary integral row operation on A is equivalent to multiplying on the left of A by a matrix $\begin{bmatrix} 1 & k \\ 0 & 1 \end{bmatrix}$ or $\begin{bmatrix} 1 & 0 \\ k & 1 \end{bmatrix}$ for some integer k. So the sequence of this operation multiplies A on the left by a matrix $B = \begin{bmatrix} s & t \\ u & v \end{bmatrix}$ with det $B = sv - ut = 1$, where s, t, u, v are integers. Therefore

$$BA = \begin{bmatrix} s & t \\ u & v \end{bmatrix} \begin{bmatrix} a & 1 & 0 \\ b & 0 & 1 \end{bmatrix} = \begin{bmatrix} sa + tb & s & t \\ ua + vb & u & v \end{bmatrix}.$$

Now if $ua + vb = 0$, then $(ua' + vb')g = 0$ where $g = $ GCD(a, b) and $ua' + vb' = 0$. Since GCD$(a', b') = 1$, we obtain $u = db', v = -da'$, for some integer d. Now $sv - ut = 1 \implies s(-da') - db't = 1 \implies d(sa' + tb') = -1 \implies d = \pm 1, sa' + tb' = \mp 1 \implies sa + tb = g(sa' + tb') = \pm g$. Similarly if $sa + tb = 0$, then $ua + vb = \pm g$. These matrices show that the GCD approach to matrices is a proof. (See also Silvester [11]).

Example 139. Find the GCD(611, 235) and a particular solution of the integral linear combination by matrix method. Compare the matrix method with the procedure in Example 43.

Solution. As described in this example, we first write a 2 × 3 matrix augmented with the identity in the last two columns.

$$A = \begin{bmatrix} 611 & 1 & 0 \\ 235 & 0 & 1 \end{bmatrix}.$$

Using (−2) times Row 2 and adding it to Row 1 we have

$$B = \begin{bmatrix} 141 & 1 & -2 \\ 235 & 0 & 1 \end{bmatrix}.$$

Subtracting Row 1 of B from Row 2 of B gives

$$C = \begin{bmatrix} 141 & 1 & -2 \\ 94 & -1 & 3 \end{bmatrix}.$$

Subtracting Row 2 from Row 1 of C yields

$$D = \begin{bmatrix} 47 & 2 & -5 \\ 94 & -1 & 3 \end{bmatrix}.$$

From which we obtain

$$E = \begin{bmatrix} 47 & 2 & -5 \\ 0 & -5 & 13 \end{bmatrix}.$$

The GCD(611, 235) is the (1, 1) entry of E and the coefficients of the integral combination of 611 and 235 are the entries in the (1, 2) and (1, 3) spots. More specifically:

$$\text{GCD}(611, 235) = 47 \text{ and } 47 = 2(611) - 5(235)$$

as before.   //

Remark. The matrix method is quite useful because it both finds the GCD and simultaneously expresses the GCD as an integral linear combination.

Exercise 91. Use the matrix method to find the GCD(24, 116) and write GCD(24, 116) as an integer linear combination of 24 and 116.

This matrix method can also be applied to find the particular solution of linear Diophantine equation of two variables.

Example 140. Find the particular solution of the integer solution of $16x+39y=10$.

Solution. First we set up the matrix as in Example 139

$$\begin{bmatrix} 16 & 1 & 0 \\ 39 & 0 & 1 \end{bmatrix}.$$

Add -2 times row 1 to row 2, we have

$$\begin{bmatrix} 16 & 1 & 0 \\ 7 & -2 & 1 \end{bmatrix}.$$

Add -2 times row 2 to row 1, we have

$$\begin{bmatrix} 2 & 5 & -2 \\ 7 & -2 & 1 \end{bmatrix}.$$

Add -3 times row 1 to row 2, we have

$$\begin{bmatrix} 2 & 5 & -2 \\ 1 & -17 & 7 \end{bmatrix}.$$

Add -2 times row 2 to row 1, we have

$$\begin{bmatrix} 0 & 39 & -16 \\ 1 & -17 & 7 \end{bmatrix}.$$

Multiply row 2 by 10, we obtain

$$\begin{bmatrix} 0 & 39 & -16 \\ 10 & -170 & 70 \end{bmatrix}.$$

Thus the particular solution is $x = -170$ and $y = 70$.   //

The next step is to apply the matrix method to find the greatest common divisor of polynomials, $d(x) = \text{GCD}(a(x), b(x))$. It also is important to write $d(x)$ as the linear combination of $a(x)$ and $b(x)$ using polynomials. Here are a number of examples.

Example 141. Use the matrix method to find $g(x) = \text{GCD}(x^5 + 6x^2 - 49x + 42, x^4 + x^3 - 9x^2 - 3x + 18)$.

Solution.

$$\begin{bmatrix} x^5 + 6x^2 - 49x + 42 & 1 & 0 \\ x^4 + x^3 - 9x^2 - 3x + 18 & 0 & 1 \end{bmatrix} \xrightarrow{R_1 = r_1 - xr_2}$$

$$\begin{bmatrix} -x^4 + 9x^3 + 9x^2 - 67x + 42 & 1 & -x \\ x^4 + x^3 - 9x^2 - 3x + 18 & 0 & 1 \end{bmatrix} \xrightarrow{R_1 = r_1 + r_2}$$

$$\begin{bmatrix} 10x^3 - 70x + 60 & 1 & -x + 1 \\ x^4 + x^3 - 9x^2 - 3x + 18 & 0 & 1 \end{bmatrix} \xrightarrow{R_2 = 10r_2 - xr_1}$$

$$\begin{bmatrix} 10x^3 - 70x + 60 & 1 & -x + 1 \\ 10x^3 - 20x^2 - 90x + 180 & -x & x^2 - x + 10 \end{bmatrix} \xrightarrow{R_2 = r_2 - r_1}$$

$$\begin{bmatrix} 10x^3 - 70x + 60 & 1 & -x + 1 \\ -20x^2 - 20x + 120 & -x - 1 & x^2 + 9 \end{bmatrix} \xrightarrow{R_1 = 2r_1 + xr_2}$$

$$\begin{bmatrix} -20x^2 - 20x + 120 & -x^2 - x + 2 & x^3 + 7x + 2 \\ -20x^2 - 20x + 120 & -x - 1 & x^2 + 9 \end{bmatrix} \begin{array}{l} \scriptstyle R_1 = r_1 - r_2 \\ \scriptstyle R_2 = -r_2 \end{array}$$

$$\begin{bmatrix} 0 & -x^2 + 3 & x^3 - x^2 + 7x - 7 \\ 20x^2 + 20x - 120 & x + 1 & -x^2 - 9 \end{bmatrix}$$

So $g(x) = 20x^2 + 20x - 120$ and $(x + 1)(x^5 + 6x^2 - 49x + 42) + (-x^2 - 9)(x^4 + x^3 - 9x^2 - 3x + 18) = 20x^2 + 20x - 120$.   //

Example 142. Use the matrix method to find $g(x) = \text{GCD}(x^4 + 3x - 2, 2x^3 + x^2 + x + 6)$.

Solution.

$$\begin{bmatrix} x^4 + 3x - 2 & 1 & 0 \\ 2x^3 + x^2 + x + 6 & 0 & 1 \end{bmatrix} \xrightarrow{R_1 = 2r_1 - xr_2}$$

$$\begin{bmatrix} -x^3 - x^2 - 4 & 2 & -x \\ 2x^3 + x^2 + x + 6 & 0 & 1 \end{bmatrix} \xrightarrow{R_2 = 2r_1 + r_2}$$

$$\begin{bmatrix} -x^3 - x^2 - 4 & 2 & -x \\ -x^2 + x - 2 & 4 & -2x + 1 \end{bmatrix} \xrightarrow{R_1 = r_1 - xr_2}$$

$$\begin{bmatrix} -2x^2 + 2x - 4 & -4x + 2 & 2x^2 - 2x \\ -x^2 + x - 2 & 4 & -2x + 1 \end{bmatrix} \begin{array}{l} \scriptstyle R_1 = -\frac{1}{2}r_1 \\ \scriptstyle R_2 = -r_2 \end{array}$$

$$\begin{bmatrix} x^2 - x + 2 & 2x - 1 & -x^2 + x \\ x^2 - x + 2 & -4 & 2x - 1 \end{bmatrix} \xrightarrow{R_1 = r_1 - r_2}$$

$$\begin{bmatrix} 0 & 2x + 3 & -x^2 - x - 1 \\ x^2 - x + 2 & -4 & 2x - 1 \end{bmatrix}$$

Then $g(x) = x^2 - x + 2$ and $(-4)(x^4 + 3x - 2) + (2x - 1)(2x^3 + x^2 + x + 6) = x^2 - x + 2$.   //

Example 143. Use the matrix method to find $g(x) = \mathrm{GCD}(x^4 + x^3 + 2x^2 + x + 1, x^3 - 1)$.

Solution.

$$\begin{bmatrix} x^4 + x^3 + 2x^2 + x + 1 & 1 & 0 \\ x^3 - 1 & 0 & 1 \end{bmatrix} \xrightarrow{R_1 = r_1 - xr_2}$$

$$\begin{bmatrix} x^3 + 2x^2 + 2x + 1 & 1 & -x \\ x^3 - 1 & 0 & 1 \end{bmatrix} \xrightarrow{R_1 = r_1 - r_2}$$

$$\begin{bmatrix} 2x^2 + 2x + 2 & 1 & -x - 1 \\ x^3 - 1 & 0 & 1 \end{bmatrix} \xrightarrow{R_2 = 2r_2 - xr_1}$$

$$\begin{bmatrix} 2x^2 + 2x + 2 & 1 & -x - 1 \\ -2x^2 - 2x - 2 & -x & x^2 + x + 2 \end{bmatrix} \xrightarrow{R_2 = r_1 + r_2}$$

$$\begin{bmatrix} 2x^2 + 2x + 2 & 1 & -x - 1 \\ 0 & -x + 1 & x^2 + 1 \end{bmatrix}$$

So $g(x) = 2x^2 + 2x + 2$ and $(1)(x^4 + x^3 + 2x^2 + x + 1) + (-x - 1)(x^3 - 1) = 2x^2 + 2x + 2$.   //

Exercise 92. Write $d(x) = \mathrm{GCD}(a(x), b(x)) = p(x)a(x) + q(x)b(x)$ where $a(x) = 7x^3 + 6x^2 - 8x + 4$ and $b(x) = x^3 + x - 2$.

The matrix method can also be used for solving $ax + by = c$. Theorem 9 (page 64) gives a necessary and sufficient condition for a solution to exist.

Example 144. Solve $172x + 20y = 1000$.

Solution. Since $\mathrm{GCD}(172, 20) = 4 \mid 1000$, it has solutions.

$$\begin{bmatrix} 172 & 1 & 0 \\ 20 & 0 & 1 \end{bmatrix} \xrightarrow{R_1 = r_1 - 8r_2} \begin{bmatrix} 12 & 1 & -1 \\ 20 & 0 & 1 \end{bmatrix} \xrightarrow{R_2 = r_2 - r_1}$$

$$\begin{bmatrix} 12 & 1 & -8 \\ 8 & -1 & 9 \end{bmatrix} \xrightarrow{R_1 = r_1 - r_2} \begin{bmatrix} 4 & 2 & -17 \\ 8 & -1 & 9 \end{bmatrix} \xrightarrow{R_2 = r_2 - 2r_1} \begin{bmatrix} 4 & 2 & -17 \\ 0 & -5 & 43 \end{bmatrix}$$

So $4 = (2)(172) + (-17)(20) \Rightarrow 1000 = 250(4) = 250((2)(172) + (-17)(20)) = 500(172) - 4250(20)$. So one solution is $x_0 = 500$ and $y_0 = -4250$. Therefore, by Theorem 26, all solutions are of the form

$$x = 500 + \left(\frac{20}{4}\right)t = 500 + 5t, \quad y = -4250 - \left(\frac{172}{4}\right)t = -4250 - 43t.$$

//

Exercise 93. Determine all integer solutions to the following Diophantine equations by matrix methods.

(a) $56x + 72y = 40$
(b) $24x + 138y = 18$
(c) $221x - 354y = 11$

Problem Set 6.5:

1. Using the Matrix Method of Example 139, find the greatest common divisor of the two given integers. Then write the gcd as a linear combination of the two integers.

   (a) 260 and 90
   (b) 1785 and 340
   (c) −616 and 286

2. Determine the gcd for each pair of numbers from Problem 1 using the Euclidean Algorithm from Chap. 2.

3. Using your work from Problem 2, determine a solution to the integral linear combination problem for each pair of numbers from Problem 1 using Tableau method.

4. Given integers 297 and 693,

   (a) Determine GCD(297, 693) using the Euclidean Algorithm and then solve the integral linear combination problem using back substitution.
   (b) Determine GCD(297, 693) and solve the integral linear combination problem using the Matrix Method.
   (c) Write a paragraph describing how these procedures work and the correspondence between the repeated division in the Euclidean Algorithm and the row operations in the Matrix Method.

5. Find $d = \text{GCD}(a, b)$ and integer x and y such that $d = ax + by$ by using matrix method

   (a) $a = 52, b = 96,$
   (b) $a = 15, b = 50,$
   (c) $a = 5312, b = 2046.$

6. Find the particular solution of

   (a) $7x + 15y = 51,$
   (b) $16x + 33y = 100,$
   (c) $101x + 99y = 437.$

7. Write $d(x) = \text{GCD}(a(x), b(x)) = p(x)a(x) + q(x)b(x)$.
   $a(x) = x^4 + 3x - 2$ and $b(x) = 2x^3 + x^2 + x + 6.$

6.6 Evaluations of Determinants of 3 × 3 Matrices

Overview Section 6.6 is a review for some routine results (such as Theorems 27, 28, and 29) which you've seen already. We provide an effective way to calculate the 3 by 3 determinant.

Theorem 28. *Let*

$$A = \begin{bmatrix} a_1 & b_1 & c_1 \\ a_2 & b_2 & c_2 \\ a_3 & b_3 & c_3 \end{bmatrix}$$

If B is obtained by multiplying a row by a constant k, then $\det B = k \det A$.

Proof. Let

$$B = \begin{bmatrix} a_1 & b_1 & c_1 \\ ka_2 & kb_2 & kc_2 \\ a_3 & b_3 & c_3 \end{bmatrix},$$

then

$$\det B = \begin{vmatrix} a_1 & b_1 & c_1 \\ ka_2 & kb_2 & kc_2 \\ a_3 & b_3 & c_3 \end{vmatrix} = -ka_2 \begin{vmatrix} b_1 & c_1 \\ b_3 & c_3 \end{vmatrix} + kb_2 \begin{vmatrix} a_1 & c_1 \\ a_3 & c_3 \end{vmatrix} - kc_2 \begin{vmatrix} a_1 & b_1 \\ a_3 & b_3 \end{vmatrix}$$

$$= k \left(-a_2 \begin{vmatrix} b_1 & c_1 \\ b_3 & c_3 \end{vmatrix} + b_2 \begin{vmatrix} a_1 & c_1 \\ a_3 & c_3 \end{vmatrix} - c_2 \begin{vmatrix} a_1 & b_1 \\ a_3 & b_3 \end{vmatrix} \right)$$

$$= k \begin{vmatrix} a_1 & b_1 & c_1 \\ a_2 & b_2 & c_2 \\ a_3 & b_3 & c_3 \end{vmatrix}.$$

Similar calculation demonstrate the result when a multiplication is applied to another row. ∎

Example 145. Note that

$$\begin{vmatrix} \sqrt{2} & \sqrt{2} & \sqrt{2} \\ 2 & 1 & 5 \\ \sqrt{3} & \sqrt{3} & -\sqrt{3} \end{vmatrix} = \sqrt{6} \begin{vmatrix} 1 & 1 & 1 \\ 2 & 1 & 5 \\ 1 & 1 & -1 \end{vmatrix}.$$

Example 146. Note that

$$\begin{vmatrix} 2 & 8 & 16 \\ 3 & 24 & 30 \\ 5 & 15 & -5 \end{vmatrix} = 30 \begin{vmatrix} 1 & 4 & 8 \\ 1 & 8 & 10 \\ 1 & 3 & -1 \end{vmatrix}$$

Theorem 29. *Let*

$$A = \begin{bmatrix} a_1 & b_1 & c_1 \\ a_2 & b_2 & c_2 \\ a_3 & b_3 & c_3 \end{bmatrix}.$$

If B is obtained from A by adding a multiplication of a row of A to another row of A, then $\det B = \det A$. *The same is true when interchanging columns.*

Proof. Let

$$B = \begin{bmatrix} a_1 & b_1 & c_1 \\ a_2 + ka_3 & b_2 + kb_3 & c_2 + kc_3 \\ a_3 & b_3 & c_3 \end{bmatrix},$$

then

$$\det B = \begin{vmatrix} a_1 & b_1 & c_1 \\ a_2 + ka_3 & b_2 + kb_3 & c_2 + kc_3 \\ a_3 & b_3 & c_3 \end{vmatrix}$$

$$= -(a_2 + ka_3) \begin{vmatrix} b_1 & c_1 \\ b_3 & c_3 \end{vmatrix} + (b_2 + kb_3) \begin{vmatrix} a_1 & c_1 \\ a_3 & c_3 \end{vmatrix} - (c_2 + kc_3) \begin{vmatrix} a_1 & b_1 \\ a_3 & b_3 \end{vmatrix}$$

$$= \left(-a_2 \begin{vmatrix} b_1 & c_1 \\ b_3 & c_3 \end{vmatrix} + b_2 \begin{vmatrix} a_1 & c_1 \\ a_3 & c_3 \end{vmatrix} - c_2 \begin{vmatrix} a_1 & b_1 \\ a_3 & b_3 \end{vmatrix} \right) + k \left(-a_3 \begin{vmatrix} b_1 & c_1 \\ b_3 & c_3 \end{vmatrix} + b_3 \begin{vmatrix} a_1 & c_1 \\ a_3 & c_3 \end{vmatrix} - c_3 \begin{vmatrix} a_1 & b_1 \\ a_3 & b_3 \end{vmatrix} \right)$$

$$= \begin{vmatrix} a_1 & b_1 & c_1 \\ a_2 & b_2 & c_2 \\ a_3 & b_3 & c_3 \end{vmatrix} + k(-a_3(b_1 c_3 - b_3 c_1) + b_3(a_1 c_3 - a_3 c_1) - c_3(a_1 b_3 - a_3 b_1))$$

$$= \det A$$

∎

Example 147.

$$\begin{vmatrix} 1 & 3 & 4 \\ 2 & 1 & -3 \\ 1 & 4 & 5 \end{vmatrix} = \begin{vmatrix} 1 & 3 & 4 \\ 2 & 1 & -3 \\ 0 & 1 & 1 \end{vmatrix}$$

Solution. Adding -1 times the first row to the third row   //

Remark. The above theorem holds for determinants of all $n \times n$ matrices. Indeed, corresponding results also hold for columns and elementary column operations in all determinants because of Theorem 29. Note that elementary column operations are exactly analogous to elementary row operations.

The following three theorems are concerned with the elementary row operations and determinants of 3×3 matrices.

Theorem 30. *Let A and B be 3×3 matrices. If B is obtained from A by interchanging two rows, then* $\det B = -\det A$.

Proof. Let

$$A = \begin{bmatrix} a_1 & b_1 & c_1 \\ a_2 & b_2 & c_2 \\ a_3 & b_3 & c_3 \end{bmatrix}.$$

Without loss of generality, we may assume

$$B = \begin{bmatrix} a_2 & b_2 & c_2 \\ a_1 & b_1 & c_1 \\ a_3 & b_3 & c_3 \end{bmatrix}.$$

Then by definition,

$$\det A = a_1 b_2 c_3 + a_2 b_3 c_1 + a_3 b_1 c_2 - a_3 b_2 c_1 - a_1 b_3 c_2 - a_2 b_1 c_3$$
$$\det B = a_2 b_1 c_3 + a_1 b_3 c_2 + a_3 b_2 c_1 - a_3 b_1 c_2 - a_2 b_3 c_1 - a_1 b_2 c_3$$

Comparing $\det A$ and $\det B$, we obtain $\det B = -\det A$.   ∎

Example 148.

(a). If Row 1 and Row 2 are interchanged, then the determinant of one matrix is the negative of the other matrix.

$$\begin{vmatrix} 1/4 & 1/2 & 1 \\ 1 & 0 & 5 \\ 5 & 3 & -4 \end{vmatrix} = - \begin{vmatrix} 1 & 0 & 5 \\ 1/4 & 1/2 & 1 \\ 5 & 3 & -4 \end{vmatrix}$$

(b). If Row 2 and Row 3 are interchanged, then the determinant of one matrix is the negative of the other matrix.

$$\begin{vmatrix} 1/4 & 1/2 & 1 \\ 1 & 0 & 5 \\ 5 & 3 & -4 \end{vmatrix} = \begin{vmatrix} 1 & 0 & 5 \\ 5 & 3 & -4 \\ 1/4 & 1/2 & 1 \end{vmatrix}$$

Example 149.

$$\begin{vmatrix} 1/4 & 1/2 & 1 \\ 1 & 0 & 5 \\ 5 & 3 & -4 \end{vmatrix} = - \begin{vmatrix} 1/2 & 1/4 & 1 \\ 0 & 1 & 5 \\ 3 & 5 & -4 \end{vmatrix} \quad \text{(Column 1 and Column 2 are interchanged)}$$

Problem Set 6.6:

By using the elementary row operations and the properties of determinants, solve the following problems.

1. Find $\begin{vmatrix} b+c & a-c & a-b \\ b-c & a+c & b-a \\ c-b & c-a & a+b \end{vmatrix}$. (Answer: $8abc$)

2. If $\begin{vmatrix} l & p & u \\ m & q & v \\ n & r & t \end{vmatrix} = 8$, then find the value of $\begin{vmatrix} p & u & l \\ r & t & n \\ q & v & m \end{vmatrix} + \begin{vmatrix} l & m & n \\ p & q & r \\ u & v & t \end{vmatrix}$. (Answer: 0)

3. If $\begin{vmatrix} a & b & c \\ p & q & r \\ m & n & k \end{vmatrix} = 4$, then find the value of $\begin{vmatrix} a+3b-3c & 4b+6c & c \\ p+3q-3r & 4q+6r & r \\ m+3n-3k & 4n+6k & k \end{vmatrix}$.

4. By using the properties of determinants, simplify the following,

$$\begin{vmatrix} a & a^2 & bc \\ b & b^2 & ca \\ c & c^2 & ab \end{vmatrix} \quad \text{(Answer: } (a-b)(b-c)(c-a)abc)$$

5. By using the properties of determinants, simplify the following,

$$\begin{vmatrix} 1 & a & a^3 \\ 1 & b & b^3 \\ 1 & c & c^3 \end{vmatrix} \quad \text{(Answer: } (a+b)(b+c)(c+a)(a+b+c))$$

6. By using the properties of determinants, simplify the following,

$$\begin{vmatrix} 1 & a & a^2 \\ 1 & b & b^2 \\ 1 & c & c^2 \end{vmatrix} \quad \text{(Answer: } (a-b)(b-c)(c-a))$$

7. Show that $\begin{vmatrix} 1 & a_1 & a_2 + a_3 \\ 1 & a_2 & a_3 + a_1 \\ 1 & a_3 & a_1 + a_2 \end{vmatrix} = 0.$

6.7 Application of Determinants (Line and Area)

Overview Equations of lines and planes are often used in elementary mathematics. However, the use of matrices can result in various forms (such as two point forms for lines and areas of triangles) that highlight conceptual motivation for these topics.

Let's look at the line passing through two distinct points (x_1, y_1) and (x_2, y_2) and write $ax + by + c = 0$ as the equation of the line. Because the distinct points are on the line, (x_1, y_1) and (x_2, y_2) should satisfy this equation. Thus,

$$ax + by + c = 0$$
$$ax_1 + by_1 + c = 0$$
$$ax_2 + by_2 + c = 0.$$

We now rewrite the above linear system of linear equations as a linear system in the unknown a, b, c, obtaining

$$xa + yb + c = 0$$
$$x_1 a + y_1 b + c = 0$$
$$x_2 a + y_2 b + c = 0$$

This is a homogeneous system with unknown a, b and c. It has a nontrivial solution if and only if

$$\begin{vmatrix} x & y & 1 \\ x_1 & y_1 & 1 \\ x_2 & y_2 & 1 \end{vmatrix} = 0,$$

i.e.

$$(y_1 - y_2)x - (x_1 - x_2)y + (x_1 y_2 - y_1 x_2) = 0.$$

Theorem 31 (Two-Point Form of the Equation of a Line). *The equation of the line passing through two distinct points (x_1, y_1) and (x_2, y_2) is given by*

$$\begin{vmatrix} x & y & 1 \\ x_1 & y_1 & 1 \\ x_2 & y_2 & 1 \end{vmatrix} = 0.$$

Example 150. Find the equation of the line passing through the points $(1, 4)$ and $(-1, 5)$.

Solution. Applying the above theorem produces

$$\begin{vmatrix} x & y & 1 \\ 1 & 4 & 1 \\ -1 & 5 & 1 \end{vmatrix} = 0.$$

To evaluate this determinant, we expand by cofactors using the first row,

$$x\begin{vmatrix} 4 & 1 \\ 5 & 1 \end{vmatrix} - y\begin{vmatrix} 1 & 1 \\ -1 & 1 \end{vmatrix} + \begin{vmatrix} 1 & 4 \\ -1 & 5 \end{vmatrix} = -x - 2y + 9 = 0,$$

i.e., $x + 2y - 9 = 0$.   //

Exercise 94. Find the equation of the line passing through the points $(4, 7)$ and $(-6, 4)$.

The following corollary is the consequence of the theorem which uses the notion of collinearity

Definition. Three points or more are COLLINEAR in $\mathbb{R}^n$ if all of the points are on the same line.

Corollary 10 (Test for Collinear Points in the x, y-Plane). *Three points* (x_1, y_1), (x_2, y_2) *and* (x_3, y_3) *are collinear if and only if*

$$\begin{vmatrix} x_1 & y_1 & 1 \\ x_2 & y_2 & 1 \\ x_3 & y_3 & 1 \end{vmatrix} = 0.$$

Example 151. Determine whether points $(1, 2), (4, 5), (6, 7)$ are collinear.

Solution. Since

$$\begin{vmatrix} 1 & 2 & 1 \\ 4 & 5 & 1 \\ 6 & 7 & 1 \end{vmatrix} = 5 + 28 + 12 - (30 + 7 + 8) = 0,$$

they are collinear.   //

Exercise 95. Determine whether points $(-1, -3), (-4, -7), (2, 2)$ are collinear.

In the event that three points are not collinear, they can be viewed as the vertices of a triangle with a well defined area. In order to calculate this area directly without trigonometry, one would need to use the pythagorean theorem, the distance formula, and some geometric properties concerning triangles. On the other hand, the area can be calculated using only the information given by the coordinates of the three points using determinants.

Theorem 32. *The area of the triangle $\triangle PQR$ whose vertices are $P(x_1, y_1)$, $Q(x_2, y_2)$, $R(x_3, y_3)$ is*

$$Area = \pm\frac{1}{2}\begin{vmatrix} x_1 & y_1 & 1 \\ x_2 & y_2 & 1 \\ x_3 & y_3 & 1 \end{vmatrix},$$

where the sign $\pm$ is chosen to give a positive area.

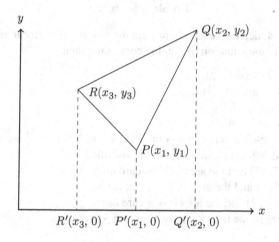

Proof. From the diagram below, we see that

Area of $\triangle PQR$ = Area of trapezoid $RR'Q'Q$ − Area of trapezoid $RR'P'P$

$\qquad\qquad$ − Area of trapezoid $PP'Q'Q$

$$= \frac{(x_2 - x_3)(y_2 + y_3)}{2} - \frac{(x_1 - x_3)(y_1 + y_3)}{2} - \frac{(x_2 - x_1)(y_1 + y_2)}{2}$$

$$= \frac{1}{2}(x_1 y_2 + x_2 y_3 + x_3 y_1 - x_1 y_3 - x_2 y_1 - x_3 y_2)$$

$$= \frac{1}{2}\begin{vmatrix} x_1 & y_1 & 1 \\ x_2 & y_2 & 1 \\ x_3 & y_3 & 1 \end{vmatrix}.$$

Note if P is above the line QR, then the area is the negative of the determinant. ∎

Example 152. Find the area of the triangle whose vertices are $(1, 1), (2, 1)$ and $(-3, 4)$.

Solution. We simply need the determinant

$$\frac{1}{2}\begin{vmatrix} 1 & 1 & 1 \\ 2 & 1 & 1 \\ -3 & 4 & 1 \end{vmatrix} = \frac{3}{2}.$$

//

Exercise 96. Find the area of the triangle whose vertices are $(-1, 0.5)$, $(2, 1)$ and $(-2, 3)$.

Problem Set 6.7:

For Problems 1–4, determine whether or not the points are collinear and in the event that they are, find the equation of the line containing them.

1. $(4, 17)$, $(6, -3)$, and $(2, 5)$.
2. $(3, 7)$, $(12, 10)$, and $(30, 16)$.
3. $(3, 12)$, $(1, 1)$, and $(2, 7)$.
4. $(0, 3)$, $(2, 0)$, and $(-4, 8)$.
   For Problems 5–7, determine a point $P(x, y)$ such that P along with the given points Q and R form a triangle with the specified area.
5. $Q(3, 8)$, $R(7, 11)$ and the area is 12 square units.
6. $Q(0, 0)$, $R(8, 0)$ and the area is 32 square units.
7. $Q(5, 7)$, $R(-1, 7)$ and the area is 6 square units.
8. What happens if one tries to apply Theorem 31 when the three points P, Q, and R are collinear?

   (a) Apply Theorem 31's area equation to the sets of collinear points from Problems 1–4.
   (b) Make a conjecture about the value of the expression:

   $$\frac{1}{2}\begin{vmatrix} x_1 & y_1 & 1 \\ x_2 & y_2 & 1 \\ x_3 & y_3 & 1 \end{vmatrix}$$

   in the event (x_1, y_1), (x_2, y_2), and (x_3, y_3) are collinear.
   (c) Draw a diagram similar to the one in the proof of Theorem 31 to support your claim and write a paragraph justifying it.

9. The following points, P, Q, and R are collinear. What is the value of a in terms of b if $P = (3, 5)$, $Q = (-1, 4)$, and $R = (a, b)$?

Selected Answers

Section 1.1

1. (a) $1, 3, 5, -9, 207$
   (b) The set $\mathcal{O}$ is the set of all integers that are one less than an even number.
   (c) No, there is no contradiction between the two definitions.
11. $a = 2, b = 2$
14. $x = 6$ or $x = -2$
17. When $n = 1$, we see that $p(1) = 1 + 1 + 17 = 19$ which is prime.

Section 1.2

2. $x = 0, 1$
4. $n = 2, 4$
5. $a = 10, b = 2$
11. $n = 18, 29$

Section 1.3

1. (a) The quotient is 3 and the remainder is 12.
   (b) $63 - 17 = 46, 46 - 17 = 29, 29 - 17 = 12$.
4. (a) $q = -3, r = 2$
   (b) You never reach -7.
   (c) You lose the negative sign.
8. 2
11. 6

© Springer International Publishing Switzerland 2015
R.S. Millman et al., *Problems and Proofs in Numbers and Algebra*,
DOI 10.1007/978-3-319-14427-6

Section 1.4

1. When $c = 0$, a must be 2, 5, or 8. When $c = 5$, a must be 0, 3, 6, or 9.

9. $a = 0$ and $b = 7$

14. $7^n = (6 + 1)^n = 6k + 1$, so $7^n - 1 \equiv 1 - 1 \equiv 0$ on division by 6.

16. 20502, 21312, 22122, 23832, 24642, 25452, 26262, 27072, 27972, 28782, 29592.

Section 1.5

1. (a) $(9A)_{12}$
   (b) $(71)_{12}$
   (c) $(101001110)_2$
   (d) $(10001001)_2$

3. (a) $(10001111)_2$
   (b) $(100011)_2$
   (c) $(1001)_2$
   (d) $(10001)_2$

Section 2.1

1. 2

4. 154

11. 7 weeks

13. (a) $(120, 10)$, $(60, 20)$, $(40, 30)$
   (b) $120, 60, 120$

Section 2.2

1. 23

8. $(11, 121)$, $(55, 77)$

Section 2.3

1. b and c

3. $x = 162$, $y = 36$ is one solution so $x = 162 + 22t$, $y = 36 + 5t$ for integer t yields all solutions.

Section 2.4

1. (a) 45
   (b) 34
   (c) 28
4. (a) False. $n = 2$
   (b) False. $n = 2$ and $m = 1$
   (c) True
12. 2^8

Section 2.5

9. $\binom{18}{14}$

Section 3.1

1. (a) $\{0\}$
   (b) $\{1, -1\}$
5. Even and odd

Section 3.2

1. (a) $5, 5$
   (b) 10
   (c) 3
1 Div. $1 = 1, 2 + 2 + 2 = 6, 3 + 3 = 6\ 4 + 4 + 4 + 4 = 16$

Section 3.3

Section 3.4

Section 4.1

1. $(x - 1)(x + 1)(5x - 2)$
10. $a = 5, b = -1$
11. -1

Section 4.2

3. (a) $q(x) = x - 1, r(x) = 3$
(b) $q(x) = 4x + 1, r(x) = 0$
5. $r(x) = x^2 + x + 1$

Section 4.3

1. $r(x) = \frac{\sqrt{5}-18}{8}$
5. $a = 3, b = 1.$
9. $r(x) = 34 - 18\sqrt{3}$
20. $a = 0, b = 2$

Section 4.4

1. (a) $q(x) = \frac{5}{2}x^2 - \frac{11}{4}x - \frac{3}{8}, r(x) = \frac{33}{8}x - \frac{37}{8}$
(b) $q(x) = x^3 + 2x^2 + \frac{1}{2}x - \frac{3}{4}, r(x) = -\frac{1}{4}x - \frac{17}{4}$
5. (a) $q(x) = -x^2 - 4x + 5, r(x) = -24x + 12$
(b) $q(x) = 3x^3 + x, r(x) = 0$

Section 5.1

1. No
2. Yes
7. $t = -35$
11. $k = -4, t = -4$

Section 5.2

2. $x = -3, 4, 5$
5. $x = -2, \frac{7}{3}$

Section 5.3

1. (a) $4 - 2x + x^2$, $x^6 + 5x^5 - 6x^4 + 8x^3 + 40x^2 - 48x$
   (b) $-6 + x + x^2$, $x^7 - 3x^5 + 6x^4 - 49x^3 + 24x^2 + 147x - 126$

Section 6.1

1. (a) Not possible.
   (b) Not possible.
   (c) $[8\,2\,12]$

5. (a) $A^{-1} = \begin{bmatrix} \frac{2}{11} & -\frac{1}{11} \\ -\frac{1}{11} & \frac{3}{11} \end{bmatrix}$

   (b) $B^{-1} = \begin{bmatrix} -\frac{1}{4} & \frac{1}{4} \\ \frac{1}{2} & \frac{1}{2} \end{bmatrix}$

   (c) None exists.

   (d) $D^{-1} = \begin{bmatrix} \frac{1}{2} & \frac{1}{2} & -\frac{1}{2} \\ -\frac{1}{2} & \frac{1}{2} & \frac{1}{2} \\ \frac{1}{2} & -\frac{1}{2} & \frac{1}{2} \end{bmatrix}$

   (e) $E^{-1} = \begin{bmatrix} -\frac{62}{5} & \frac{3}{5} & -\frac{27}{5} \\ \frac{43}{5} & -\frac{2}{5} & \frac{18}{5} \\ 2 & 0 & 1 \end{bmatrix}$

Section 6.2

2. $x = 0$, $y = -\frac{4}{3}$
3. $y = \frac{1}{3}$, $z = \frac{14}{3}$
9. $x = 1$, $y = 0$
12. $x = \frac{1}{2}\ln\frac{17}{3}$, $y = \log_2\frac{2}{3}$

Section 6.3

1. (a) $A = \begin{bmatrix} 1 & 4 & 3 \\ 5 & 9 & 14 \\ -2 & 3 & 1 \end{bmatrix}$

(b) $A = \begin{bmatrix} 2 & -3 & 5 \\ 1 & 4 & 3 \\ -2 & 3 & 1 \end{bmatrix}$

(c) It depends on the order of operations.

2. $x = \frac{12}{7}, y = \frac{15}{7}$

Section 6.4

1. $x = \frac{127}{19}, y = \frac{18}{19}, z = -\frac{49}{19}$
2. No solutions.
3. $x = -1 + \frac{1}{3}t, y = 1 + \frac{2}{3}t, z = t$

Section 6.5

1. (a) $GCD(260, 90) = 10, 10 = (-1) \cdot 260 + (3) \cdot (90)$
   (b) $GCD(1785, 340) = 85, 85 = (1) \cdot 1785 + (-5) \cdot 340$
   (c) $GCD(-616, 286) = 22, 22 = (6) \cdot (-616) + (13) \cdot 286$

Section 6.6

Section 6.7

1. No
2. Yes. $y = \frac{1}{3}x - 2$
6. $P(0, 8)$

References

1. Burton, D.: Elementary Number Theory, 7th edn. McGraw Hill, New York (2009)
2. Clark, E., Millman, R.S.: Bluma's method: a different way to solve quadratics. GCTM Reflections (Georgia Council of Teachers of Mathematics) **53**, 19–21 (2009)
3. Gallian, J., Winters, S.: Modular arithmetic in the marketplace. Am. Math. Mon. **95**, 548–551 (1988)
4. Guy, R.: Unsolved Problems in Number Theory, 2nd edn. Springer, New York (1994)
5. Honsberger, R.: Mathematical Morsels. The Dolciani Mathematical Expositions, No. 3. The Mathematical Association of America, New York (1978)
6. Long, C., DeTemple, D., Millman, R.S.: Mathematical Reasoning for Elementary Teachers, 7th edn. Addison-Wesley, Reading (2015)
7. McCrory, R.: Mathemaicians and mathematics textbooks for prospective elementary teachers. Not. Am. Math. Soc. **52**, 20–29 (2006)
8. Mollin, R.: An Introduction to Cryptography. Chapman and Hall/CRC, Boca Raton (2001)
9. Rivest, R.L., Shamir, A., Adleman, L.: A method for obtaining digital signatures and public-key cryptosystems. Commun. ACM **21**(2), 120–126 (1978)
10. Rosen, K.H.: Discrete Mathematics and Its Applications, 6th edn. McGraw-Hill, New York (2007)
11. Silvester, J.R.: A matrix method for solving linear congruences. Math. Mag. **53**(2), 90–92 (1980)

© Springer International Publishing Switzerland 2015
R.S. Millman et al., *Problems and Proofs in Numbers and Algebra*,
DOI 10.1007/978-3-319-14427-6

Printed in the United States
By Bookmasters